D1522082

International
REVIEW OF
Neurobiology

Volume 47

In Situ Hybridization Protocols for the Brain

2nd Edition

International REVIEW OF Neurobiology

Volume 47

SERIES EDITORS

RONALD J. BRADLEY

Department of Psychiatry,
Louisiana State University Medical Center
Shreveport, Louisiana, USA

R. ADRON HARRIS

Waggoner Center for Alcohol and Drug Addiction Research,
The University of Texas at Austin, Texas, USA

PETER JENNER

Division of Pharmacology and Therapeutics,
GKT School of Biomedical Sciences,
King's College, London, UK

In Situ Hybridization Protocols for the Brain

2nd Edition

EDITED BY

WILLIAM WISDEN
Department of Clinical Neurobiology
University of Heidelberg
Heidelberg, Germany

BRIAN J. MORRIS
Institute of Biomedical and Life Sciences
University of Glasgow
Glasgow, UK

ACADEMIC PRESS

An imprint of Elsevier Science

Amsterdam Boston London New York Oxford Paris
San Diego San Francisco Singapore Sydney Tokyo

This book is printed on acid-free paper.

Academic Press
An imprint of Elsevier Science
84 Theobald's Road, London WC1X 8RR
http://www.academicpress.com

Academic Press
An imprint of Elsevier Science
525 B Street, Suite 1900, San Diego, California 92101-4495, USA
http://www.academicpress.com

ISBN 0-12-366847-6

A catalogue record for this book is available from the British Library

Typeset by M Rules, London
Printed and bound in Great Britain by
MPG Books Ltd, Bodmin, Cornwall

02 03 04 05 06 07 MP 9 8 7 6 5 4 3 2 1

CONTENTS

Chapter 3
Processing Rodent Embryonic and Early Postnatal Tissue for *In Situ* Hybridization with Radiolabelled Oligonucleotides

D. J. Laurie, P. C. U. Schrotz, H. Monyer And U. Amtmann

Chapter 4
Processing Retinal Tissue for *In Situ* Hybridization

F. Müller

Chapter 5
Processing the Spinal Cord for *In Situ* Hybridization with Radiolabelled Oligonucleotides

A. Berthele and T. R. Tölle

Chapter 6
Processing Human Brain Tissue for *In Situ* Hybridization with Radiolabelled Oligonucleotides

L. F. B. Nicholson

Chapter 7
In Situ Hybridization of Astrocytes and Neurons Cultured *In Vitro*

L. Ariza-McNaughton, C. De Felipe and S. P. Hunt

Chapter 8
In Situ Hybridization on Organotypic Slice Cultures

A. Gerfin-Moser and H. Monyer

Chapter 9
Quantitative Analysis of *In Situ* Hybridization Histochemistry

A. L. Gundlach and R. D. O'Shea

PART II
Non-radioactive *In Situ* Hybridization

CHAPTER 10
Non-radioactive *In Situ* Hybridization Using Alkaline Phosphatase-labelled Oligonucleotides

S. J. AUGOOD, E. M. MCGOWAN, B. R. FINSEN, B. HEPPELMANN AND P. C. EMSON

CHAPTER 11
Combining Non-radioactive *In Situ* Hybridization with Immunohistological and Anatomical Techniques

P. WAHLE

CHAPTER 12
Non-radioactive *In Situ* Hybridization: Simplified Procedures for Use in Whole-mounts of Mouse and Chick Embryos

L. ARIZA-MCNAUGHTON AND R. KRUMLAUF

CONTRIBUTORS

U. Amtmann, Department of Clinical Neurobiology, University Heidelberg, Im Neuenheimer Feld 364, 69120 Heidelberg, Germany

Ariza-McNaughton, Vertebrate Development Laboratory Cancer Research UK, 44 Lincoln's Inn Fields, London WC2A 3PX, UK

S. J. Augood, Neurology Research, Massachusetts General Hospital, Charlestown, MA 02129, USA

A. Berthele, Department of Neurology, Technical University Munich, Moehlstr. 28, Germany

P. Emson, Laboratory of Molecular Neuroscience, The Babraham Institute, Cambridge CB2 4AT, UK

C. De Filipe, Instituto de Neurosciencias, University Miguel Hernandez, San Juan, E-03550, Alicante, Spain

B. R. Finsen, Anatomy and Neurobiology, University of Southern Denmark – Odense University, Winsløwparken 21, DK-5000 Odense C, Denmark

A. Gerfin-Moser, Schweizerische Multiple Sklerose Gesellschaft, Brinerstrasse 1, Postfach, 8036 Zurich, Switzerland

A. L. Gundlach, Howard Florey Institute of Experimental Physiology and Medicine and Department of Medicine, Austin and Repatriation Medical Centre, The University of Melbourne, Victoria 310, Australia

B. Heppelmann, Physiologisches Institut, Universität Würzburg, Röntgenring 9, D-97070 Würzburg, Germany

S. P. Hunt, Department of Anatomy and Developmental Biology, University College London, Gower Street, London WC1E 6BT, UK

A. Jones, MRC Laboratory of Molecular Biology, MRC Centre, Hills Road, Cambridge CB2 2QH, UK

R. Krumlauf, Stowers Institute for Medical Research, 1000 East 50th Street, Kansas City, Missouri 64110, USA

D. J. Laurie, DRA Oncology, Novartis Pharma Inc, CH-4002 Basel, Switzerland

E. M. McGowan, Birdsall Building, Mayo Clinic, Jacksonville, Florida, USA

H. Monyer, Department of Clinical Neurobiology, University Heidelberg, Im Neuenheimer Feld 364, 69120 Heidelberg, Germany

B. J. Morris, Division of Neuroscience and Biomedical Systems, Institute of Biomedical and Life Sciences, University of Glasgow, Glasgow G12 8QQ, UK

F. Müller, Institut für Biologische Informationsverarbeitung 1, Forschungszentrum Jülich GmbH, D-52425 Jülich, Germany

L. F. B. Nicholson, Department of Anatomy, School of Medicine, University of Auckland, Private Bag 92019, Auckland, New Zealand

V. Revilla, MRC Laboratory of Molecular Biology, MRC Centre, Hills Road, Cambridge CB2 2QH, UK

R. D. O'Shea, Department of Pharmacology, Monash University, Victoria, Australia

P. C. U. Schrotz, Department of Neurobiology, University Heidelberg, Im Neuenheimer Feld 364, 69120 Heidelberg, Germany

T. R. Tölle, Department of Neurology, Technical University Munich, Moehlstr. 28, Germany

P. Wahle, AG Entwicklungsneurobiologie, Fakultät für Biologie, Ruhr-Universität, ND 6/72, D-44780 Bochum, Germany

W. Wisden, Department of Clinical Neurobiology, University of Heidelberg, Im Neuenheimer Feld 364, 69120 Heidelberg, Germany

ACKNOWLEDGEMENTS

The editorial staff at Academic Press in London have shown enduring patience and support. Dr Sarah Stafford commissioned the second edition; it was subsequently nurtured and gently prompted by the Life & Biomedical Sciences Editor, Lisa Tickner and her editorial assistant, Wendy Gray. They did an excellent job in making sure the book appeared. Simon Crump (Academic Press) provided excellent support with the proofs. We thank our colleagues and friends for their willingness to revamp their chapters, or write new ones.

Bill Wisden
Brian Morris
May 2002

HOW TO USE THIS BOOK

New gene sequence, new exon sequence, or faced with a whole new gene family? Nothing known about the expression? (Chapter 1) For vertebrate brains, radioactive methods should be the first choice if you have a new gene and you want to find out where it is expressed. Non-radioactive methods are less sensitive, and you will either miss the expression of your gene in certain cell types where it in fact is expressed, or at the other extreme, it is easier to get non-specific hybridization, and you will infer that your gene is expressed in more places than it actually is. Radioactive oligonucleotides are your best bet. This is a standardized and sensitive method. Follow Chapter 1, and its variations described in Chapters 3–8.

Need to quantify changes in gene expression between brain regions or between experimental conditions? The best approach is radioactive-labelled probes (Chapters 1 and 9).

You wish to know if your gene is co-expressed with another mRNA or protein (Chapters 10, 11 and 12). You already know in which cell types your gene is expressed. Your gene is expressed abundantly. Non-radioactive methods are needed. You can either use an alkaline phosphatase-conjugated oligonucleotide and co-hybridize this with a ^{35}S-labelled oligonucleotide (Chapter 10), or you can use digoxigenin-labelled cRNA probes (Chapter 11). cRNA probes are more sensitive; oligonucleotides are more specific. Both can be combined with antibody staining.

You want to look at gene expression in very young embryos (less than E10.5). Whole mounts with digoxigenin-labelled cRNA probes are the best way (Chapter 12).

Need further information? If at any point you need more background on techniques or recipes, consult the 3rd edition of "Molecular Cloning: A Laboratory Manual" (Sambrook, J and Russell, D. W., 2001, Cold Spring Harbor Laboratory Press, New York).

W. Wisden
B. J. Morris

INTRODUCTION: STUDYING GENE EXPRESSION IN NEURAL TISSUES BY *IN SITU* HYBRIDIZATION

In this genomic era everybody wants to know where his or her favourite gene or gene family is expressed. This book aims to explain the *in situ* hybridization (ISH) technique for people who want to do it. We concentrate on the vertebrate brain because this is our own research area. The many people doing ISH on invertebrates have their own successful methods, mostly based on whole-mount non-radioactive methods (Tautz and Pfeifle, 1989; Hauptmann, 1999), but these are not usually useful for looking at gene expression in large tissue masses, such as a vertebrate brain.

ISH detects the sites in a tissue where a particular mRNA is expressed. The high degree of anatomical resolution makes ISH invaluable in tissues such as the brain, where the cell types expressing a given gene can be marked out from the mosaic of cells constituting neuronal tissue. The principle is simple. A nucleic acid probe, tagged with either radiolabelled nucleotides for autoradiographic detection, or other molecules allowing colorimetric detection, is hybridized directly to a tissue section, where it recognizes its cognate mRNA and forms a probe–mRNA duplex. After excess probe is washed away, the section is either exposed to photographic film or emulsion for autoradiography, or processed histochemically to reveal the sites on the section where the particular mRNA is expressed. Under the correct conditions, the labelled "antisense" nucleic acid probe will hybridize only to the target mRNA, which has the complementary sequence.

ISH was first used to localize genes on chromosomes and then adapted for detection of intracellular RNAs (Harrison *et al.*, 1973; Brahic and Haase, 1978). In the early 1980s, the first modern ISH experiments applied to the vertebrate brain studied peptide gene expression in the pituitary and hypothalamus (Hudson *et al.*, 1981; Gee and Roberts, 1983; Gee *et al.*, 1983). The 1980s saw a revolution in and popularization of molecular biology methods, and the description of many mRNA expression patterns in the mammalian brain (e.g. Branks and Wilson, 1986; Morris *et al.*, 1986,

1988; Shivers *et al.*, 1986a,b; Young *et al.*, 1986a, b; Goedert and Hunt, 1987; Rogers and Hunt, 1987; Uhl, 1987; Voigt and Uhl, 1988; Wisden *et al.*, 1988, 1991, Emson, 1993). During this period, Young and colleagues and Watson and colleagues popularized a particular type of ISH method using synthetic oligodeoxyribonucleotides tailed with ^{35}S (Lewis *et al.*, 1985, 1988; Young *et al.*, 1986a,b; Young, 1989). Our book, "*In Situ* Hybridization Protocols for the Brain", largely based on this oligonucleotide method, was first published in 1994. Today the importance of ISH remains undiminished. ISH studies have become more widespread: indeed, the majority of laboratories doing molecular neuroscience research will employ ISH at some point. The information arriving from the genome sequencing programmes can only increase the need. Automated or semi-automated ISH methods will surely appear.

When the 1st edition was published there were many different protocols for radioactive ISH; and of course, there still are. Plurality is good: different types of probe, different labelling techniques, different ways of processing the sections and hybridizing them. But any person wishing to use the technique for the first time has great difficulty in identifying which method to follow. How do you choose? We say choose ours! The advantages of the oligonucleotide-based method described in Chapter 1 remain as relevant as ever.

1. *Simplicity*: The protocol described in Chapter 1 is reliable, sensitive and simple. The researchers who visit us to learn this technique are always pleasantly surprised to find how simple it is, and have set it up quickly in their own laboratories.
2. *Specificity*: Alternative splicing of genes expressed in the brain is the rule rather than the exception; this means that probes employed for ISH may need to be specific for splice variants. Oligonucleotides inherently have considerably increased specificity as compared with longer forms of probe, such as cRNA probes. In addition, the treatments required to ensure specific hybridization of cRNA probes tend to introduce variability into the strength of the hybridization signal. This gives oligonucleotide probes a distinct advantage for monitoring changes in mRNA expression in discrete brain regions.
3. *High throughput*: Large numbers of oligonucleotide probes can be labelled and hybridized at the same time. This is much less work than doing the same thing with cRNA probes. This should have great utility for analysing the expression pattern of potential mRNAs (including splice variants) derived from the genome projects.

In Part I, chapters 1–9 are linked, all using the same basic method of radiolabelled oligonucleotides. This basic method is described in detail in Chapter 1; Revilla and Jones provide a description of the best ways to collect cryostat sections of brain tissue (Chapter 2); Laurie *et al.* cover embryonic and postnatal tissue (Chapter 3); in Chapter 4, Müller describes procedures needed for working with retina; Berthele and Tölle (Chapter 5) describe the application of the method to spinal cord. Over the next few years, the application of ISH to human post-mortem tissue will yield important information on altered gene expression associated with neurological disease; hence Nicholson's contribution (Chapter 6). The method works well on cultured neurons and glia: Ariza-McNaughton *et al.* (Chapter 7), and Gerfin-Moser and Monyer (Chapter 8) cover adaptations for dispersed and organotypic cultures respectively. The quantification of ISH autoradiographs and emulsions is covered by Gundlach and O'Shea (Chapter 9). In addition, throughout the book we have tried to provide an honest account of the method – it doesn't always work; Chapter 1 therefore provides a troubleshooting section.

Part II covers non-radioactive ISH methods, which have the advantages of rapidity, high anatomical resolution and the ability to double- or triple-label cells. These methods need more careful application than the "bulk-standard" radioactive oligonucleotide method; everything with non-radioactive probes should be case-by-case. For vertebrate brains, radioactive methods should be the first choice if you have a new gene and you want to find out where in the brain it is expressed. For this, use the Chapter 1 method; non-radioactive methods are less sensitive, and you will either miss the expression of your gene in certain cell types where it in fact is expressed, or at the other extreme, it is easier to get non-specific hybridization, and you will get the impression that your gene expressed in more places than it actually is. With the exception of very tiny embryos, non-radioactive methods are best applied when you already know where the gene is expressed and you need to know further information such as: "Is the gene co-expressed with something else?" For the co-localization of mRNA and protein (or mRNA and mRNA) within the same cell, non-radioactive ISH has advantages. Emson's laboratory has been one of the pioneers of non-radioactive ISH methods using conjugated oligonucleotides (Emson *et al.*, 1989; Emson, 1993; Chapter 10). This allows co-localization of two different mRNAs by cohybridizing a ^{35}S-labelled oligonucleotide and one conjugated with alkaline phosphatase. The advantage of the Augood *et al.* (Chapter 10) method is that it is compatible with the Chapter 1 method. Another approach employs the "digoxigenin system" with cRNA probes (Kessler *et al.*, 1990). This important method is detailed in conjunction with immunohistochemistry and tract tracing on adult brains and in cell culture; and the co-localization of two mRNAs in the same cell (Chapter 11). For very young (E10.5 or less) embryos, digoxigenin-labelled cRNA probes hybridized to whole-mount embryo preparations are the best way to localize gene expression – whole mounts are much better than trying to section such tiny embryos (Chapter 12).

It is our experience that ISH techniques can be used in any laboratory. This book has been designed with enough detail so that someone without any previous experience of the techniques can follow the recipes and expect to get a good result.

W. Wisden, Heidelberg, May 2002
B. J. Morris, Glasgow, May 2002

References

Brahic, M. and Haase, A. T. (1978). *Proc. Natl Acad. Sci. USA* **75,** 6125–6129.

Branks, P. L. and Wilson, M. C. (1986). *Brain Res.* **387,** 1–16.

Emson, P. C. (1993). *Trends Neurosci.* **16,** 9–16.

Emson, P. C., Arai, H., Agrawal, S., Christodoulou, C., and Gait, M. J. (1989). *Methods Enzymol.* **168,** 753–761.

Gee, C. E. and Roberts, J. L. (1983). *DNA* **2,** 157–163

Gee, C. E., Chen, C. L., Roberts, J. L., Thompson, R., and Watson, S. J. (1983). *Nature* **306,** 374–376.

Goedert, M. and Hunt, S. P. (1987). *Neuroscience* **22,** 983–992.

Harrison, P. R., Conkie, D., Paul, J., and Jones, K. (1973). *FEBS Lett.* **32,** 109–112.

Hauptmann, G. (1999). *Dev. Genes Evol.* **209,** 317–321.

Hudson, P., Penschow, J., Shine, J., Ryan, G., Niall, H., and Coghlan, J. (1981). *Endocrinology* **108,** 353–356.

Kessler, C., Höltke, H. J., Seibl, R., Burg, J., and Mühlegger, K. (1990). B*iol. Chem. Hoppe Seyler* **371,** 917–927.

Lewis, M. E., Sherman, T. G., and Watson, S. J. (1985). *Peptides* **6** (Suppl. 2), 75.

Lewis, M. E., Krause, R. G., and Roberts-Lewis, J.M. (1988). *Synapse* **2,** 308.

Morris, B. J., Haarmann, I., Kempter, B., Hollt, V., and Herz, A. (1986). *Neurosci. Lett.* **69,** 104–108.

Morris, B. J., Feasey, K. J., ten Bruggencate, G., Herz, A., and Hollt, V. (1988). *Proc. Natl Acad. Sci. USA* **85,** 3226–3230.

Rogers, J. H. and Hunt, S. P. (1987). *Neuroscience* **23,** 343–361.

Shivers, B. D., Harlan, R. E., Romano, G. J., Howells, R. D., and Pfaff, D. W. (1986a), and *Proc. Natl Acad. Sci. USA* **83,** 6221–6225.

Shivers, B. D., Schachter, B. S., and Pfaff, D. W. (1986b). *Methods Enzymol.* **124,** 497–510.

Tautz, D. and Pfeifle, C. (1989). *Chromosoma* **98,** 81–85.

Uhl, G. R. (Ed.) (1987). "*In situ* Hybridization in Brain". Plenum Press, New York.

Voigt, M. M. and Uhl, G. R. (1988). *Brain Res.* **464,** 247–253.

Wisden, W., Morris, B. J., Darlison, M. G., Hunt, S. P., and Barnard, E. A. (1988). *Neuron* **1,** 937–947.

Wisden, W., Morris, B. J., and Hunt S. P. (1991). *In* "Molecular Neurobiology: A Practical Approach" (J. Chad and H. Wheal, Eds), pp. 205–225. Oxford University Press/IRL Press, Oxford.
Young, W. S., 3rd (1989). *Methods Enzymol.* **168,** 702–710.
Young, W. S., 3rd, Bonner, T. I., and Brann, M. R. (1986a). *Proc. Natl Acad. Sci. USA* **83,** 9827–9831.
Young, W. S., 3rd, Mezey, E., and Siegel, R. E. (1986b). *Mol. Brain Res.* **1,** 231–241.

Part I

In situ Hybridization with Radiolabelled Oligonucleotides

CHAPTER 1

IN SITU HYBRIDIZATION WITH OLIGONUCLEOTIDE PROBES

W. Wisden* and B. J. Morris**

*Department of Clinical Neurobiology, University of Heidelberg, Im Neuenheimer Feld 364, Heidelberg, Germany

**Division of Neuroscience and Biomedical Systems, Institute of Biomedical and Life Sciences, University of Glasgow, Glasgow, Scotland, UK

INTERNATIONAL REVIEW OF NEUROBIOLOGY, VOL. 47
ISSN/02 $00.00

1.1 Introduction: mRNA Hybridization Using Oligodeoxyribonucleotide Probes

> Science is most at home attacking problems that require technique rather than insight. By technique we mean the systematic application of a sequential procedure – a recipe. John D. Barrow. *Theories of Everything.*

> There is never any justification for things being complex when they could be simple. Edward De Bono. *Simplicity.*

In situ hybridization (ISH) is an important method for tracing the regional and cellular sites of gene expression (mRNA distribution) within a tissue. This method comes into its own when applied to the intermingled cell types of the brain. ISH allows the particular cell type in which the mRNA is contained to be marked out from others (Branks and Wilson, 1986). The technique is simple. A nucleic acid probe, tagged with either radiolabelled nucleotides or molecules allowing colorimetric/light detection, is applied to a tissue section. The probe searches out and finds (hybridizes to) its corresponding mRNA in the cells and forms a probe–mRNA double helix. After the excess probe has been washed away, the section is either exposed for autoradiography or processed histochemically to reveal the sites on the section targeted by the probe.

1.2 ISH with Oligonucleotides

ISH to brain tissue with oligodeoxyribonucleotides (oligonucleotides) was developed and popularized by Young and collaborators, Lewis and collaborators, and others (Lewis *et al.*, 1985, 1988; Morris *et al.*, 1986; Young *et*

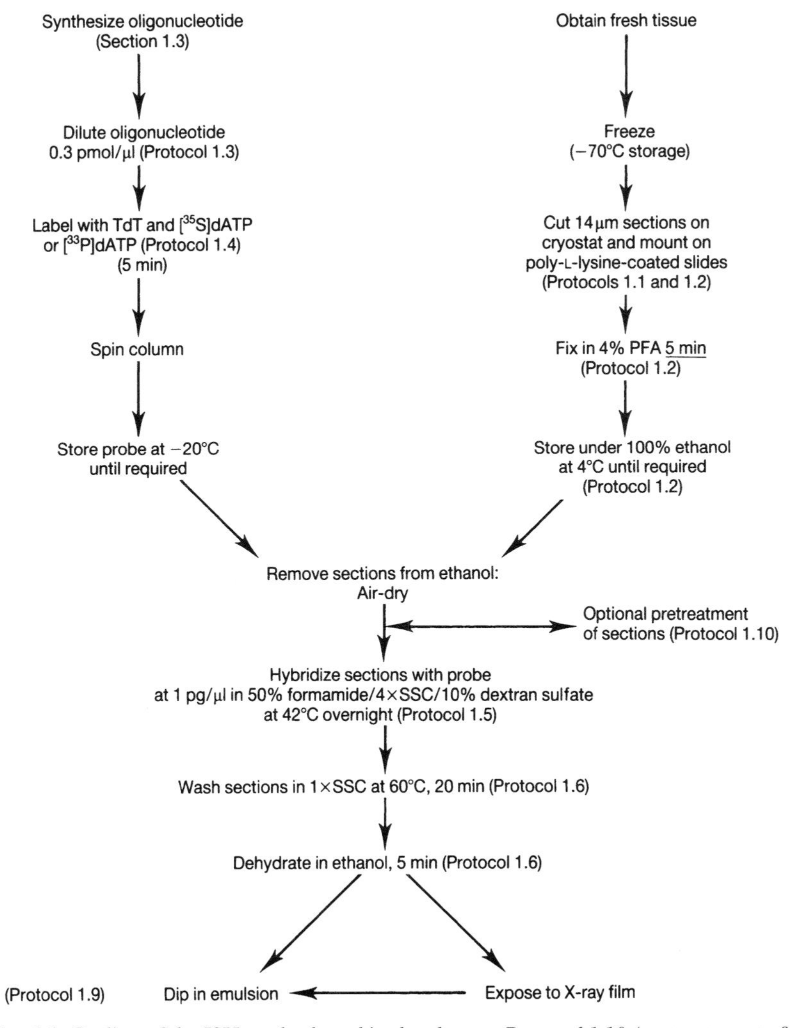

FIG. 1.1. Outline of the ISH method used in the chapter. Protocol 1.10 (pretreatment of sections) is only used for special tissues, e.g. embryos, adrenal gland and retina. The spin column step is Protocol 1.4.1.

al., 1986a,b; Young, 1989; reviewed, Wisden *et al.*, 1991; Emson, 1993; Pratt and Kokaia, 1994). ISH with radiolabelled oligonucleotides is straightforward. There are many variations, but the protocol we describe here is the simplest. The scheme is summarized in Fig. 1.1.

Slide-mounted sections are cut from unfixed frozen tissue on a cryostat.

Sections are then fixed with paraformaldehyde and stored in ethanol at 4°C until required (Section 1.5.3). Oligonucleotides (Section 1.6) are radiolabelled ("tailed") using terminal deoxyribonucleotide transferase and "hot" deoxyadenosinetriphosphate (usually [α^{35}S]dATP or [α^{33}P]dATP, and occasionally with [α^{32}P]dATP; see Chapter 9, Section 9.2.2, Choice of iostope). The labelling reaction is described in Section 1.7. Labelled probes are diluted in hybridization buffer (Section 1.8), the probe–hybridization buffer mix is applied to a brain section (Section 1.8), and hybridized overnight. The next day excess probe is washed off, and after dehydration the sections can be exposed to either X-ray film to produce a global image (Section 1.9.3) or dipped in photographic emulsion for cellular resolution (Section 1.9.4). Controls are discussed in Section 1.10 and presentation in Section 1.12. Even though this is a clear and simple method, ISH does not always give perfect results. Troubleshooting is covered in Section 1.11.

Chapters 2–9 are integrated with Chapter 1, and form Part I of this book. These chapters all describe using the same oligonucleotide hybridization method, but adapted to particular tissues. How the cryostat works, and how to make adult brain sections is described in Chapter 2; in Chapter 3 for embryos; in Chapter 4 for retina; in Chapter 5 for spinal cord; and in chapter 6 for human brain. Neuronal and glial cell cultures (dispersed and organotypic) can also be hybridized with radiolabelled oligonucleotides – this is covered in Chapters 7 and 8. Chapter 9 discusses the quantification of ISH data. Localizing two different mRNAs in the same cell is accomplished by co-hybridizing a ^{35}S-labelled oligonucleotide with an oligonucleotide tagged with alkaline phosphatase (Chapter 10, Section 10.7).

1.3 Advantages of Using Oligonucleotides for ISH

ISH with oligonucleotides needs less "hands-on" time than other ISH methods. First, oligonucleotides are ordered to specified sequences. So subcloning or RT-PCR steps to produce, for example, templates for cRNA or cDNA probes are not needed. Second, many aspects of the ISH method add time and expense. We have systematically removed these factors. Many ISH protocols have pretreatments: proteinase K to increase probe access to mRNA, acetic anhydride to cover the section with negative charge to repel nucleic acids, and chloroform treatment to remove lipids, which could also absorb nucleic acids non-specifically. These pretreatments are generally not needed for vertebrate brain sections hybridized with radiolabelled oligonucleotide probes. Similarly, prehybridization steps, whereby the section is pre-incubated with all of the hybridization buffer components except probe to absorb out non-specific binding sites, are superfluous and can be eliminated. Many components in traditional hybridization buffers (e.g. salmon sperm DNA) are unnecessary (see Section 1.8.3 and Lewis *et al.*, 1988). However, certain types of tissue section (such as those derived from embryo, retina, adrenal gland) benefit from acetic anhydride/chloroform treatment (see Protocol 1.10 Optional Pretreatment of Sections prior to Hybridization) and/or maximalist buffer for spinal cord (see Chapter 5).

1.3.1 High Throughput

Large numbers of oligonucleotides (e.g. up to 20 at one time) and sections (e.g. 100) can be handled at once. The method is well suited to large-scale studies that, for example, compare the expression of all members of a gene family or map the expression of novel gene sets (Marvanová et al 2002). Large-scale mapping with cRNA probes is of course possible; this was successfully done recently for genes with enriched expression in the amygdala (Zirlinger *et al.*, 2001), but we think it would have been easier with oligonucleotides.

1.3.2 Standardization and Reproducibility.

All oligonucleotides of an approximately equal length require the same hybridization conditions. This is not true for cRNA probes: their hybridizations need fine-tuning for each new probe fragment. In contrast to cRNA probes, oligonucleotides give a highly reproducible hybridization signal with minimal intra-sample variation. This makes ISH with oligonucleotide probes ideal for detecting changes in mRNA levels between tissue samples, i.e. quantitative studies (Johnston and Morris, 1994, 1995; Thomas *et al.*, 1994, 1996; Morris, 1995, 1997; Roberts *et al.*, 1998; Hall *et al.*, 2001; Thomas and Everitt, 2001 – see also Chapter 9).

1.3.3 Specificity

Gene Families

The oligonucleotide method works well for mapping gene families (Morris *et al.*, 1990; Wisden *et al.*, 1990, 1992, 2000; Malosio *et al.*, 1991; Laurie *et al.*, 1992a, b; Persohn *et al.*, 1992; Tölle *et al.*, 1993; Wisden and Seeburg, 1993; Thomas *et al.*, 1994, 1996; Berthele *et al.*, 1998; Paterlini *et al.*, 2000; Sinkkonen *et al.*, 2000; Campos *et al.*, 2001). (See Figs 1.2 and 1.3.)

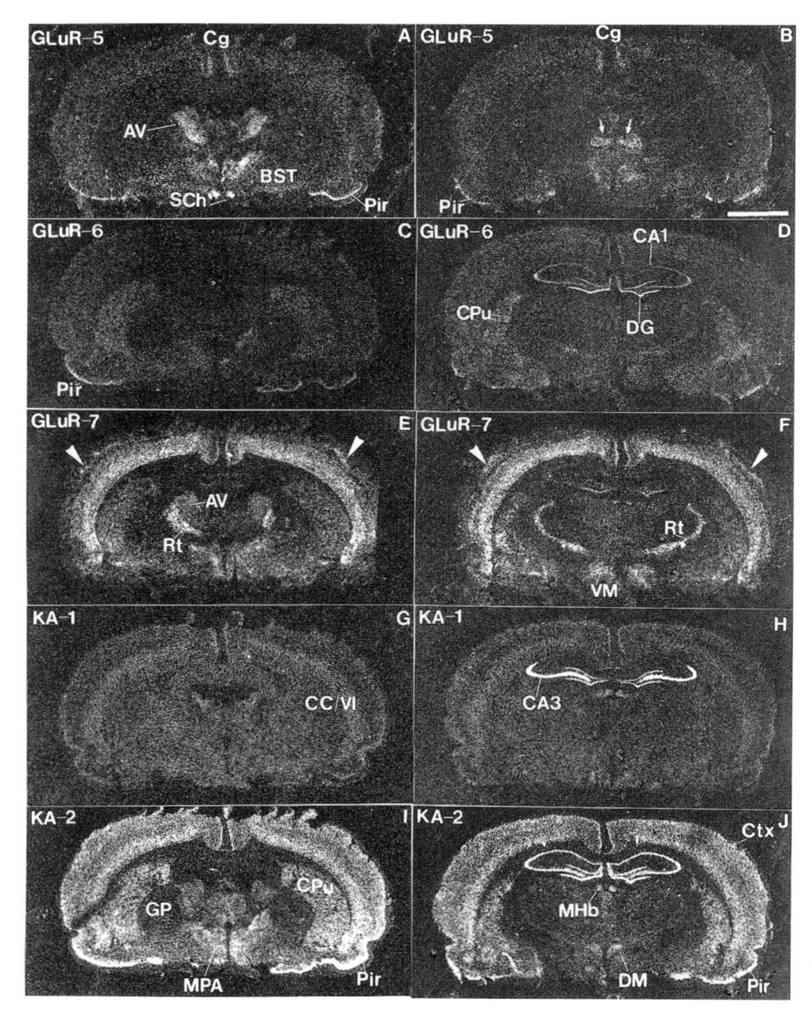

FIG. 1.2. Mapping the expression of gene families by ISH with gene-specific oligonucleotides I. This figure illustrates the expression of genes encoding the kainate receptor subunit mRNAs in the adult rat brain. Coronal sections were hybridized with ^{35}S-labelled oligonucleotide probes specific for each gene (GluR5–7, KA-1, KA-2); sections were exposed to Kodak XAR-5 film, and the images printed onto light-sensitive paper. White areas correspond to the presence of mRNA. Arrowheads in E and F mark neocortical layer III cells expressing the GluR7 gene. AV, anteroventral thalamic nucleus; BST, bed nucleus stria terminalis; CC, corpus callosum white matter tract; Cg, cingulate cortex; CPu, caudate putamen; DG, dentate granule cells; DM, dorsomedial hypothalamic nucleus; GP, globus pallidus; MPA, medial preoptic area; Pir, piriform cortex; Rt, reticular thalamic nucleus; SCh, suprachiasmatic nucleus; Scale bar 3.2 mm (Reproduced with permission from Wisden and Seeburg, 1993. ©1993 by the Society for Neuroscience.)

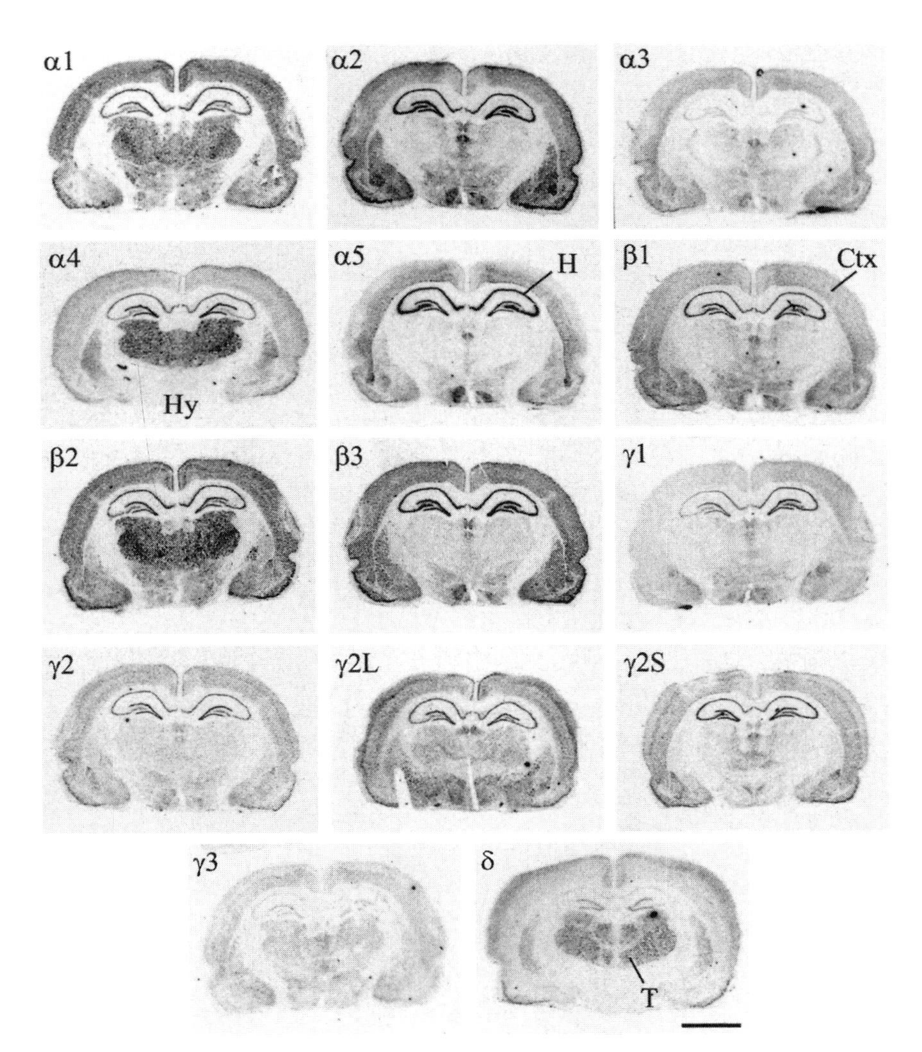

FIG. 1.3. Mapping the expression of gene families by ISH with gene-specific oligonucleotides II. This figure illustrates the expression of genes encoding the $GABA_A$ receptor subunit mRNAs in the adult rat brain. Coronal sections were hybridized with ^{35}S-labelled oligonucleotide probes specific for each gene (α1–α5, β1–β3, γ1–γ3, δ) or splice forms (γ2S and γ2L); sections were exposed to Kodak BioMax film, digitally scanned, and then manipulated in Adobe Photoshop. Black areas mark the presence of mRNA. Ctx, neocortex; H, hippocampus, Hy, hypothalamus; T, thalamus. Scale bar 2 mm (D. Merlo and W. Wisden, unpublished and Campos *et al.*, 2001; probe sequences from Wisden *et al.*, 1992, and Miralles *et al.*, 1994).

There is a risk that longer cRNA probes will cross-hybridize to related family members.

mRNA Splice Variants

Oligonucleotides can precisely distinguish the expression patterns of mRNAs differing in specific exons, or between genes differing in only small regions of nucleotide sequence: for example, splice variants of AMPA-type glutamate receptor mRNAs (Sommer *et al.*, 1990); splice variants of $GABA_A$ receptor mRNAs (see Fig. 1.3; Glencorse *et al.*, 1992; Miralles *et al.*, 1994; Jones *et al.*, 1996); splice variants of the NMDA receptor subunit genes (Laurie and Seeburg, 1994; Laurie *et al.*, 1995); and splice forms of metabotropic glutamate receptor genes (Berthele *et al.*, 1998, 1999). Oligonucleotide probes would be ideal for mapping brain expression patterns of the many exons of the cadherin-related neuronal receptor genes (Wu and Maniatis, 1999; Sugino *et al.*, 2000). Oligonucleotides can also be built to hybridize specifically to intronic sequences, so allowing the localization of pre-mRNA in brain (Higuchi *et al.*, 2000). Building oligonucleotides to detect mRNA splice variants is described in Section 1.6.4.

Exon Specificity

This is related to mRNA splicing. ISH with exon-specific oligonucleotide probes can look at expression levels of particular exons in mutant mice. This can confirm that the targeted exon is no longer expressed (Hormuzdi *et al.*, 2001), or that a particular exon has a different expression level in a mutated gene (Mulhardt *et al.*, 1994; Higuchi *et al.*, 2000).

1.3.4 Automation

The oligonucleotide approach will be best for automating the ISH method. The expression of the 30 000–40 000 or so genes in the mouse genome could be feasibly mapped by systematic ISH with oligonucleotides, For example, two oligonucleotides could be built from each EST fragment or predicted gene (RIKEN Genome Exploration Research Group Phase II Team and the FANTOM Consortium, 2001). Construction of databases to analyse the data will be crucial; for example, the Edinburgh Mouse Atlas Project (Davidson and Baldock, 2001; Davidson *et al.*, 2001).

1.4 Problems

Usually the ISH method works well. However, three things can frustrate you: cutting good sections (see Chapter 2, Section 2.5); probe labelling to the correct specific activity (Section 1.7.2); and artefacts from emulsion dipping (Protocol 1.9). More rarely, there are also artefacts produced on the X-ray film images (see Fig. 1.18 and Section 1.11).

1.5 Preparation of Sections

1.5.1 Dissecting Brain Samples

Intact RNA is not needed for ISH. There is time to dissect a fresh brain and prepare it for cutting. In vertebrate brains, mRNA is stable for a long period after death (Pittius *et al.*, 1985; Uhl *et al.*, 1985; Ross *et al.*, 1992; Gilmore *et al.*, 1993; Harrison *et al.*, 1995; Kingsbury *et al.*,1995). Even after a 48-hour post-mortem interval, rat brain RNA is essentially undegraded; and even 96 hours after death, 40% of the RNA is still present (Harrison *et al.*, 1995). These long RNA survival times are probably because cellular compartments must first disintegrate before RNases, normally in separate cellular compartments, can get to their RNA substrate. For human post-mortem brain, tissue pH is an indicator of mRNA preservation; acidic tissue (pH 5.8–6.3) is more likely to contain degraded mRNA, possibly through the activation of acid ribonucleases (Barton *et al.*, 1993; Harrison *et al.*, 1995; Kingsbury *et al.*, 1995). (See also Chapter 6 on ISH to human brain sections.)

In summary, even if, after the death of the animal, a brain has been 2 days at room temperature, ISH should still work (Harrison *et al.*, 1995). Berthele and Tölle caution that spinal cord has a high concentration of RNases, and so this tissue should be dissected rapidly (see Chapter 5).

1.5.2 Freezing and Fixing the Brain

We use non-perfused tissue for ISH. After removal of the brain, it is slowly frozen by placing it on a flat square of aluminum foil on a metal plate; this is then placed on top of powdered dry ice. Once it has been completely frozen (usually within 5–10 min), the brain can be wrapped in parafilm (to prevent freeze-drying – freeze-drying makes it difficult to cut as the tissue tends to crumble on the knife), placed in a polypropylene screw-cap tube (e.g. 50 ml Falcon) or a sealable plastic freezer bag and transferred to –70°C storage (for months to years) until needed for cutting. In the medium term, RNA is stable in tissue stored at –70°C; no loss of mRNA occurred in tissue stored for 33 months at –70°C (Harrison *et al.*, 1995). After freezing the tissue must not at any point be allowed to thaw: this may crack open the intracellular contents and allow RNase access to RNA, as well as destroying much of the cellular morphology. (The brain is not snap-frozen by dipping it into methanol/dry ice or liquid nitrogen as this shatters the tissue and makes it difficult to cut good sections on the cryostat.)

On the cutting day, the frozen material is transferred from the –70°C freezer to the –20°C cryostat chamber to equilibrate (for at least 1 h) prior to cutting (see Chapter 2). Sections are usually 15 μm thick (they can be thinner or thicker if you want, depending on the application). When using ^{32}P-labelled probes (see Section 1.7.8), thicker sections may give a higher signal as they contain more mRNA, and the β-electrons are energetic enough to escape from the deeper parts of the tissue. There is no point in using thicker sections for ^{35}S or ^{33}P to increase sensitivity, as the β-electrons will not escape from the tissue's depths. Alternatively, extra thin (5 μm) sections can be used for serially sectioning through large cells (e.g. motor neurons of the vertebrate spinal cord). This approach has been used to deduce co-expression of different mRNAs in the same cell (see Chapter 5).

On the day of cutting, fresh 4% paraformaldehyde (Protocol 1.2) and poly-L-lysine-coated slides (Protocol 1.1) are prepared. Older solutions of paraformaldehyde may contain oxidation products, which could damage the RNA in the tissue section. Poly-L-lysine is the most effective substrate for ensuring section adhesion to slides for ISH. Sections never come off even at high wash temperatures; probably, the formaldehyde physically cross-links the tissue section to the lysine residues on the glass. The slide coating takes 10 min for 50 slides (see Protocol 1.1), and once dry, they are ready to use. Coated slides can be stored for an extended period (many months) at 4°C.

For the solutions for fixing sections onto the slides, and rinsing and alcohol dehydration we use the staining troughs shown in Fig. 1.4.

1.5.3 Advantages of Long-term Storage of Sections Under Ethanol

After cryostat cutting, most protocols recommend storing sections desiccated at –70°C. In our experience this is inconvenient for two reasons: RNA stability and organization of samples.

- ***RNA stability***: Retention of mRNA in sections is poor if they are

PROTOCOL 1.1 PREPARATION OF POLY-L-LYSINE-COATED SLIDES

1. Dissolve 25 mg poly-L-lysine hydrobromide[a] in 5 ml of DEPC-treated water. Store as 1-ml aliquots at –20°C.
2. Thaw out one 1-ml tube of the 5 mg/ml aliquots of poly-L-lysine, and dilute it to 50 ml with DEPC-treated water, to give a 0.01% solution. Transfer this solution to a sterile 50-ml disposable plastic Petri dish.
3. Dip the baked slides[b] individually in the poly-L-lysine solution (immerse each slide completely), and allow slides to air-dry standing upright in a rack. Slides can be stored for a few weeks at 4°C without any loss of adhesive properties.

a Poly-L-lysine hydrobromide, MW 350 000 (Sigma, P-1524).
b A packet of standard 75 mm × 26 mm (e.g. BDH/Merck) microscope slides is wrapped in aluminium foil and baked (to sterilise them) for 4 h to overnight at 180°C. Slides are allowed to cool to room temperature before use. It is best to use slides that have a frosted edge so that they can be labelled easily with pencil. This is more convenient than using a diamond pen on a non-frosted slide, and just as durable.

freeze–thawed more than once. Once thawed, sections should not be refrozen.

- ***Organization:*** Consider the following scenario. After six months of storing slide boxes at –70°C, you come to do an experiment. Even if you are good at labelling samples, it is natural to be unsure either how many slides, or exactly what sections are in each box. You cannot see how many sections are in each box without thawing the box, and you probably don't want to use all your slides in a single experiment – so some will either have to be refrozen, which is something to be avoided, or discarded.

Storing sections in ethanol is more versatile than putting them in the freezer, principally because slides can be quickly removed from the ethanol to see what they are, and quickly replaced before they dry. You can see exactly how many sections/slides you have, and what the sections look like; they can easily be reorganized and regrouped, and large libraries of slides/sections can be maintained (Fig. 1.5).

Sections stored under ethanol are stable for long periods (but make sure the storage box lids are tightly sealed to prevent evaporation). Sections stored for over 5 years in this way have shown no loss of signal. Storing sections in ethanol is equivalent to storing pure RNA as precipitates in ethanol – it is the best way for long-term RNA storage. Under ethanol, the dangers of natural hydrolysis or enzyme attack of RNA in the tissue section are minimized. Once the sections have been removed from ethanol, they are dry within minutes and are ready to be hybridized. In contrast, thawing

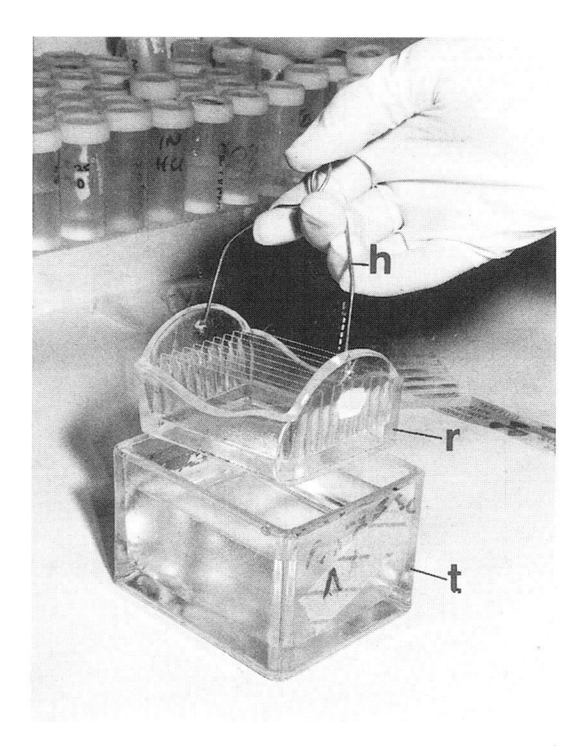

FIG. 1.4. The staining trough (t), slide rack (r) and wire handle (h) used for transferring the sections between different solutions (available from NEOLAB, Heidelberg, Germany, or the BDH/Merck Laboratory Supplies Catalogue).

out sections from the –70°C freezer is time consuming and requires more rigid planning before starting an experiment.

1.5.4 STERILITY

ISH does not require obsessive precautions to prevent ribonuclease contamination from outside sources. There is probably more RNase in the tissue than in external sources; but mRNA in a fixed tissue section is partially protected against RNase. There is no need to use sterile equipment to remove the brain. However, if only for psychological security, it is advisable to take the following two precautions: use autoclaved diethylpyrocarbonate (DEPC)-treated water to make up the solutions used before the hybridization step and for the hybridization buffer itself; wear gloves when handling reagents/slides before the hybridization step (see How to win the battle with

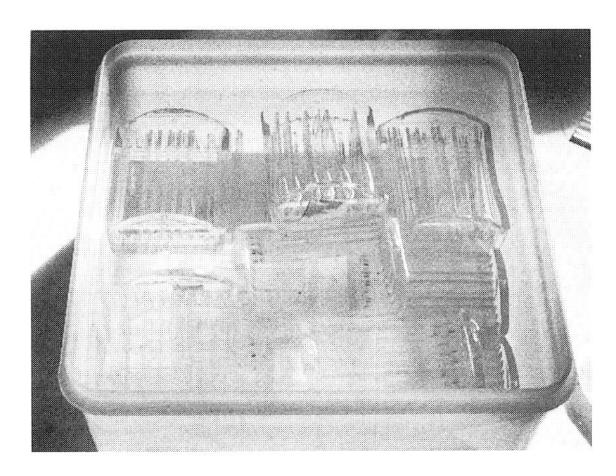

FIG. 1.5. The storage of racks of sections under ethanol. A plastic/Tupperware box such as those used for home freezing (e.g. "Seal fresh" Pizza Storers; Stewart) is filled with a couple of litres of absolute ethanol. Racks of slides are placed into the box, and then sealed with the lid (not shown). IMPORTANT: Ethanol-filled boxes must be stored in the cold room. Do NOT store large quantities of ethanol in refrigerators – if the thermostat sparks, it could cause an explosion.

RNase in Sambrook and Russell, 2001, volume 2, section 7.82). It is unlikely that any small contaminating quantities of RNase will survive the paraformaldehyde used for fixation, be active in the 100% ethanol used to store the sections, or be active in the hybridization buffer containing 50% formamide at elevated temperatures (see Chapter 11, Section 11.5.5 for more discussion on this). Baking destroys RNases. If the glassware for fixing and rinsing the sections prior to ethanol storage is thoroughly cleaned (e.g. in a dishwasher), it is not necessary to bake glassware. However, baked microscope slides are used for the cryostat sections (Protocol 1.1). After the hybridization step, there is no need to use sterile washing and staining solutions: DNA/RNA hybrids resist RNase attack.

1.6 Oligonucleotide Probes: Synthesis and Design

1.6.1 Synthesis and Purification

Oligonucleotides are made on programmable machines (for background information, see Sambrook and Russell, 2001, volume 2, panel 10.42). Competitive offers for oligonucleotide synthesis can be found on the Web

PROTOCOL 1.2 PREPARATION AND FIXATION OF SECTIONS

1. Cut 12–15 μm sections on a cryostat at –20°C. Thaw mount the sections onto poly-L-lysine-coated slides.[a] Allow sections to dry at room temperature for half an hour to several hours. For organizational purposes, slides can be consecutively numbered on their frosted area with pencil markings. Such pencil markings are effectively permanent. They do not come off in the ethanol storage or during hybridization/washing steps.
2. Prepare a 4% solution of paraformaldehyde (PFA) as follows. Transfer 40 g paraformaldehyde[b] into 1 litre 1 × PBS (or in proportion for smaller volumes). The PBS does not need to be DEPC-treated, as the paraformaldehyde will inactivate any RNases. Heat the milky white suspension with continuous stirring until it reaches around 60–65°C, when it will dissolve. Do not heat beyond the point where the solution becomes clear, and in particular, do not allow to boil. Do the whole procedure in fume hood. Chill the solution in an ice/water bath. It is probably best to make the PFA solution fresh each day.
3. Transfer a rack of dry sections into the ice-cold 4% PFA. Leave for 5 min[c].
4. Transfer sections into 1 × PBS[d] for several minutes. The exact time is not critical.
5. Transfer sections into 70% ethanol[e] for several minutes.
6. Transfer the sections into the storage box[f] containing 95% ethanol. Store at 4°C in cold room until required. The storage of large quantities of 95% ethanol in refrigerators is not recommended because of the risk of ignition by sparking from thermostats. Note also that industrial grade ethanol (containing 5% methanol) works just as well as the more expensive pure grade.

Glass staining troughs that hold 250 ml of solution (size: 93 × 75 × 55 mm internal) and matching glass racks with grooves (allowing transfer from one solution to another without removing the slides) are obtained from Arnold R. Horwell Ltd (UK) or Merck Ltd/BDH. Stainless steel wire handles for the glass racks are also available from BDH/Merck. These are listed in the Histology section of the Merck Laboratory Supplies Catalogue. Staining troughs and handles are illustrated in Figs 1.4 and 1.5.

[a] See Protocol 1.1 for the preparation of poly-L-lysine-coated slides.

[b] Paraformaldehyde (powder) is general-purpose reagent grade, e.g. Merck/BDH.

[c] We use 19 slides/glass rack and 250 ml of solution in continental staining troughs. See Section 1.5 for details of glassware.

[d] 10 × PBS is 1.3M NaCl, 70 mM Na_2HPO_4, 30 mM NaH_2PO_4. To make 1 litre 10 × PBS, dissolve 75.79 g NaCl, 9.93 g Na_2HPO_4 anhydrous and 4.68 g $NaH_2PO_4.2H_2O$. Filter, DEPC treat and autoclave. 2 × PBS and 1 × PBS are prepared by diluting the 10× stock with sterile water.

[e] Diluted from 100% ethanol with DEPC-treated water.

[f] Large plastic pizza boxes used for home freezing make convienient storage tanks for large numbers of rack. Alternatively, sections can be stored in the continental troughs.

PROTOCOL 1.3 DILUTION OF OLIGONUCLEOTIDES FOR ISH

Typically, oligonucleotides arrive from our Applied Biosystems machines dissolved in TE (10 mM Tris pH 7.0, 1 mM EDTA) buffer at around an average concentration of 1000 pmol/µl (i.e. 1 mM). A working stock solution used for TdT labelling is 0.3 pmol/µl (Protocol 1.4).

1. Take 10 µl of concentrated stock and dilute to 100 µl in sterile H_2O (Dilution 1). (This (1/10) dilution can be kept as a permanent reserve of oligonucleotide stored at –20°C.) Also store the original synthesis stock at –20°C.
2. Take 5 µl of Dilution 1 (the 1/10 stock) and dilute to 1000 µl in H_2O (Dilution 2).
3. Measure the optical absorbance of the *total* 1000 µl of Dilution 2 at 260 nm (OD_{260}) with reference to sterile water as a blank.
   For an oligonucleotide, assume 1.0 OD_{260} unit corresponds to 20 µg/ml (Sambrook and Russell, 2001).
   For a 45mer oligonucleotide, 0.3 pmol/µl is equivalent to 5 ng/µl (assuming average molecular weight of one nucleotide is 330).
4. Dilute the appropriate amount of oligonucleotide from Dilution 1 (the 1/10 dilution) into 1000 µl of sterile H_2O to give a working stock of 5 ng/µl (for a 45mer).

This 5 ng/µl stock is stored at –20°C and can be freeze–thawed (*ad infinitum*) as required.
Specific example:
For a particular 45mer.
5 µl of Dilution 1 in 1000 µl of water gives OD_{260} reading of 0.2.

i.e. 5 µl of Dilution 1 contains $0.2 \times 20 = 4$ µg oligonucleotide.
Thus 1 µl of Dilution 1 contains 0.8 µg oligonucleotide.

To prepare 1 ml of 5 ng/µl (0.3 pmol/µl) labelling stock, add 6.25 µl Dilution 1 into 1000 µl of H_2O.

(e.g. http://genedetect.com/insitu.htm or http://www.genosys.co.uk/). Consult the Nature Biotechnology Directory and Buyers' Guide Online (http://guide.nature.com/). Entering the keyword "oligonucleotide" into the search box currently gives 138 companies/university departments supplying custom oligonucleotides. There is a huge choice.

Only small amounts of oligonucleotide are needed for radioactive ISH experiments; 0.3 picomoles (approximately 5 ng of a 45mer) of oligonucleotide are enough to hybridize 50 large horizontal rat brain sections. One synthesis run of a particular oligonucleotide should be a lifetime's supply. We find it unnecessary to purify oligonucleotides. The concentrated oligonucleotide solution from the machine is usually in the millimolar range, i.e. 1000 pmol/μl; the solution may look dirty and sometimes even yellow, but when serially diluted to the low concentrations used for labelling (0.3 pmol/μl, see Protocol 1.3), any contaminants from the synthesis reaction are not a problem. Alternatively, commercial oligonucleotides may arrive freeze-dried with the weight and moles of oligonucleotide specified. In this case they can simply be dissolved in TE buffer to a stock solution of 1000 pmol/μl (mix extremely thoroughly, as freeze-dried oligonucleotides can resist resuspension) and stored at –20°C long term. An aliquot from this concentrated stock is used to prepare the 0.3 pmol/μl labelling stock.

Oligonucleotides are stable pieces of DNA: the end result of most DNase reactions is to cut the DNA into oligonucleotides rather than individual nucleotides. In the short term, oligonucleotide stocks can be stored at room temperature. The stocks are frozen to prevent evaporation and the slow sticking/absorption of the oligonucleotide to the walls of the tube.

A potential advantage in purifying an oligonucleotide (by acrylamide gel electrophoresis) is that it removes the shorter polymer products present from the synthesizing reaction (see Sambrook and Russell, 2001, volume 2, chapter 10, protocol 1, "Purification of synthetic oligonucleotides by polyacrylamide gel electrophoresis"). In practice, these shorter forms do not interfere with the ISH, so we never gel-purify the probes.

1.6.2 Ordering Oligonucleotides

How do I write out my probe? The sequence of your probe must be antisense, so that it can hybridize to the sense mRNA. Published annotated DNA sequences are usually given as the sense (coding or plus) strand. Probes built to detect mRNA are made as complementary sequence to this, and with the 5′ to 3′ polarity opposite to that of the sense strand. For example, if the sense strand sequence is 5′-GAATTCCCGGG ... 3′, then the sequence of the probe will be 5′-CCCGGGAATTC ... 3′.

1.6.3 Oligonucleotide Probe Design

Using the protocol in this chapter, oligonucleotides can be anywhere in length from 36mers (Sommer *et al.*, 1990; Monyer *et al.*, 1991; Laurie and Seeburg, 1994) to 60mers (Wisden *et al.*, 1990) using similar hybridization and wash conditions. We routinely use 40 or 45mers with no difference in results; for abundant mRNAs, 36mers can be hybridized stringently under the same conditions used for 45mers (Sommer *et al.*, 1990). Here are some guidelines for designing an oligonucleotide for hybridization studies (Lewis *et al.*, 1988; Erdtmann-Vourliotis *et al.*, 1999). If an oligonucleotide has too many A or T residues it may hybridize less efficiently. AT and AU (uracil in RNA) base pairs form by two non-valent contacts and are less stable than GC pairs, which form three. Unless there is no other choice, it is not a good idea to build an oligonucleotide that is excessively GC rich – such a probe could give more non-specific binding. The probe should be designed such that the GC/AT ratio is between 50 and 65%. Certain oligonucleotides form hairpin structures and may hybridize with themselves internally, at least to some extent, rather than with the target mRNA. Because less probe is available, this reduces detection sensitivity. In practice, it is best to be empirical and to use several oligonucleotides that hybridize to different parts of the mRNA (see Section 1.10 on controls). It might be useful to run the intended sequence of your oligonucleotide against the EMBL/Genbank database (BLAST search) to see if it is likely to hybridize to anything else (http://www.ncbi.nlm.nih.gov/BLAST/). However, assuming that the probes do not recognize unique mRNA splice variants (Section 1.6.4), multiple oligonucleotides giving the same autoradiographic pattern serve as the best internal control (Section 1.10).

Computer programs can predict internal hairpins, melting temperature and hybridization temperature. We have nothing against these programs, but we seldom use them. We choose the sequences by eye, using the AT and GC criteria (see above). There is a failure rate: some oligonucleotides (maybe one in ten) do not work. So if you want to be more systematic, Erdtmann-Vourliotis *et al.* (1999) have looked at ways of optimizing probe design.

1.6.4 Oligonucleotide Probes for Detecting mRNA Splice Forms

We consider two types of RNA splicing. First, use of alternative exons, where each is mutually exclusive in the final mRNA, or the exon is a simple clean insertion of extra sequence into the mRNA, e.g. the flip and flop exons of the AMPA-type glutamate receptor subunits or the N1 exon of the NMDA receptor NR1 gene (Fig. 1.6; Sommer *et al.*, 1990; Laurie and Seeburg, 1994). In this case, oligonucleotides are designed within the inserted exon sequence (Fig. 1.6). Provided that there are at least 35 nucleotides of unique sequence specific for each exon, it is clear how to write out a specific probe. The number "35" is the lower limit of oligonucleotide length for use with the 50% formamide/4 × SSC hybridization stringency (Section 1.8.3). Intron-specific probes to study the expression of pre-mRNA are designed similarly (Higuchi *et al.*, 2000).

Second, and more challenging, is the detection of mRNAs with deleted

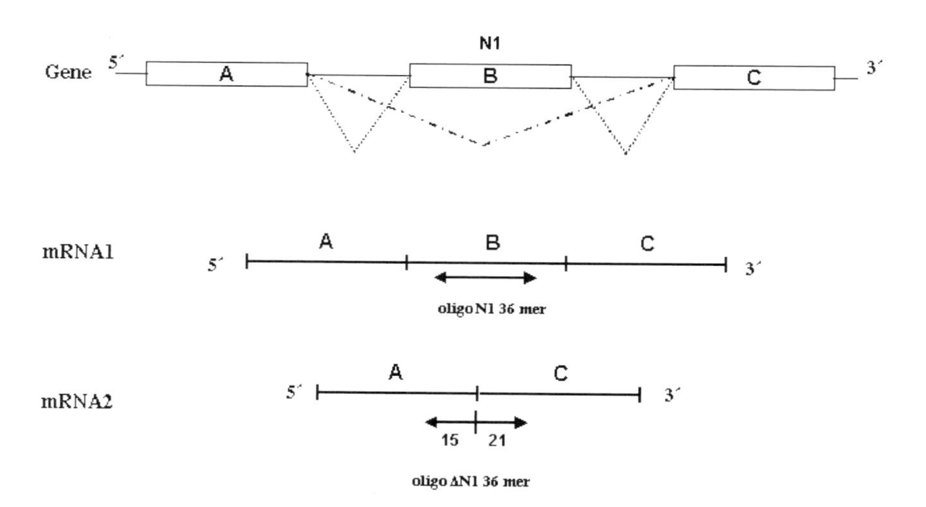

FIG. 1.6. Designing oligonucleotides to detect alternative RNA splicing. A fragment of the NMDA NR1 receptor subunit gene is shown; three exons, A, B and C are marked as boxes; the hashed and dotted lines show alternative splicing pathways. Exon B, the N1 exon, can be either present (mRNA1) or absent (mRNA2). To detect mRNA1 by ISH, oligonucleotide N1 (a 36mer) is built within exon N1; to detect mRNA2, but *not* mRNA1, an oligonucleotide (DN1 36mer) bridging the deletion between A and C is built. The arms of the oligonucleotide (lengths 15 and 21 nucleotides) are too short to hybridize to the RNA domains A and C individually; a stable probe–mRNA hybrid can form only if domains A and C are directly adjacent (see Section 1.6.4). (Adapted from Laurie and Seeburg, 1994. ©1994 by the Society for Neuroscience.)

exons. An excellent example is a study by Laurie and Seeburg (1994). They describe the brain distribution of the eight mRNA splice variants of the NMDA receptor NR1 subunit gene (Laurie and Seeburg, 1994). To detect an mRNA lacking an exon (e.g. to detect NR1 mRNA lacking the N1 cassette), an oligonucleotide spanning the deletion point is built, with, for example, 19–25 base arms each side of the exon deletion (Fig. 1.6). Under stringent conditions of hybridization (50% formamide/4 × SSC at 42°C, Section 1.8.3), these flanking arms are too short to hybridize to mRNAs that contain the N1 cassette; the only possible oligonucleotide–mRNA hybrids will be those with no N1 cassette sequence (Fig. 1.6). The specificity of the oligonucleotide probes can be checked by dotting the corresponding cDNAs onto membranes and hybridizing them with the ^{32}P-labelled (kinased) oligonucleotides specific for each splice form (Laurie and Seeburg, 1994). To reproduce the ISH conditions more closely, the probes can be 3′-tailed (Section 1.7.1) with non-radioactive dATP before 5′-labelling with [γ^{32}P]ATP. Cross-hybridization is further reduced by the 3′ polyadenosine tail (Laurie and Seeburg, 1994).

1.7 Probe Labelling with Terminal Deoxynucleotidyl Transferase (TdT)

1.7.1 Principles

ISH works best with a low probe concentration to reduce non-specific binding and hence requires high-specific-activity probes for optimal results. For this reason, labelling oligonucleotides on their 5′ end with only one radioactive atom ("kinasing") does not give probes with high enough specific activity. Instead, oligonucleotides are labelled with terminal deoxynucleotidyl transferase (often called terminal transferase, TdT), which puts multiple radioactive residues onto the end of the oligonucleotide. TdT is a DNA polymerase found in pre-lymphocytes of the developing immune system. It synthesizes polydeoxyribonucleotides from deoxyribonucleotide triphosphates with the release of pyrophosphate (reviewed in Sambrook and Russell, 2001, volume 3, A4.27). TdT initiates the reaction from the free terminal 3′-hydroxyl group of single-stranded DNA, for example, oligonucleotides. The reaction is:

$$\text{single-stranded DNA}_{\text{OH}} + n\ \text{dNTP} = \text{DNA-}(_{\text{p}}\text{dN})_n + n\ \text{PPi}$$

Thus if [α^{35}S]dATP (or [α^{33}P]dATP), oligonucleotide and TdT are mixed in a buffer containing co-factors, the end result is a poly-deoxyadenylic (poly[^{35}S]dA) tail added to the 3′ end of the original oligonucleotide. The number of [^{35}S]dA residues added to the 3′ end (which is directly proportional to the probe specific activity) can be controlled by changing the molar ratio of [α^{35}S]dATP to oligonucleotide. In our hands, a 30:1 molar ratio of [α^{35}S]dATP to oligonucleotide in the labelling reaction is optimal. Longer tails give higher non-specific signals, especially over the cerebellar granule cell layer. Shorter tails require longer exposure times.

1.7.2 Basic Labelling Protocol

The TdT-labelling protocol is given in Protocol 1.4. The TdT enzyme has unusual co-factor requirements (cobalt and potassium cacodylate), and it is more convenient to use a supplier (e.g. Boehringer-Mannheim) of the TdT enzyme that also supplies the reaction buffer. Preparing the potassium cacodylate solution is tedious, but if you want to prepare your own tailing buffer then use the recipe as described in Eschenfeldt *et al.* (1987).

PROTOCOL 1.4 TERMINAL TRANSFERASE LABELLING REACTION PROTOCOL

For dilution of oligonucleotides, see Protocol 1.3.

In a standard labelling reaction, label 0.3 pmol of oligonucleotide (usually 45mers) with 10 pmol [α^{35}S]dATP or [α^{33}P]dATP, i.e. using a 30 : 1 molar ratio of isotope to oligonucleotide. This results in the addition of approximately 10–20 AMP residues to the 3′ end of the oligonucleotide (as assessed by acrylamide gel analysis). This small quantity of labelled oligonucleotide (0.3 pmol) is enough to hybridize 50 standard size microscope slides (see text).

Use terminal transferase (TdT 25 U/μl) from Boehringer-Mannheim. Reaction buffer (5×) (potassium cacodylate, 1 M, Tris–HCl, 125 mM, bovine serum albumin, 1.25 mg/ml, pH 6.6 at 25°C) and 25 mM cobalt chloride solution are supplied by the manufacturer with the enzyme. Store all components at –20°C and freeze–thaw as required.

For 10 μl reaction volume (can be scaled up to 20 μl, add more buffer but still use 1 μl of oligonucleotide and 1.5 μl of isotope, and add only 30 μl TE at the end):

1. Mix 1 μl oligonucleotide at concentration of 0.3 pmol/μl
   (note that for a 45mer, 0.3 pmol is equivalent to 5 ng/μl)
   2 μl 5 × reaction buffer (Boehringer-Mannheim)

   0.6 μl 25 mM $CoCl_2$ (Boehringer-Mannheim)

   1.5 μl [α^{35}S]dATP (1300 Ci/mmol, DuPont, NEN, NEG-034H). For ^{33}P, use 10 pmol, i.e. 1 μl of [α^{33}P]dATP

   5 μl sterile DEPC-treated H_2O
   1 μl TdT at 25 U/μl (Boehringer-Mannheim).

2. Incubate for 5–8 min at 37°C. The exact labelling time may vary slightly depending on the batch of isotope, or if using a different supplier of enzyme. Certain enzyme brands require up to 1h to achieve the same specific activity.
3. Stop the reaction by adding 40 μl TE buffer (10 mM Tris, 1 mM EDTA), or sterile water, or TNES buffer (see footnote *d*, Protocol 1.4.1); it does not matter which.
4. Apply the total 50 μl from step 3 to a Sephadex G-25 spin column (see Protocol 1.4.1) or to a commercial spin column (we use Qiaquick columns from Qiagen, according to the manufacturer's instructions). For the homemade columns, spin at 2000 rpm for 2 min. This removes unincorporated nucleotides. The volume of the column eluate should be between 40 μl and 50 μl. Count 2 μl of the eluate using liquid scintillation counting. The counts should be in the range 50 000 dpm/μl to 200 000 dpm/μl. If counts are below the 50 000 dpm/μl level, do not use the probe because

the specific activity is too low. If they are above 300 000 dpm/μl, then in our experience, some non-specific binding may occur, probably because the AMP tail is too long. Counts for ^{33}P should also be at least 100 000–200 000 dpm/μl. ^{33}P is also counted with scintillation liquid.

5. If the ^{35}S probe is OK, add 1 μl of 1M DTT to the eluate.[a] This preserves the probe from oxidation. ^{33}P requires no DTT addition. Probes are stored at –20°C and can be freeze–thawed for repeated use. We use them for up to a month after labelling.

[a] DTT (dithiothreitol). 1M stock. 3.09 g in 20 ml sterile H_2O. Store in 1-ml aliquots at –20°C. Freeze–thaw as required, but keep on ice.

Protocol 1.4.1 Spin Columns[a]

1. Plug the bottom of a 1-ml disposable syringe with a few autoclaved glass beads (see Fig. 1.7).[b]
2. Fill the syringe with Sephadex G-25 resin[c] pre-equilibrated in TNES buffer.[d] Add resin suspension by slowly dripping it into the top of the column with a pipette. Keep adding resin until the syringe is completely full.
3. Insert the syringe into a 15-ml Falcon tube (see Fig. 1.7). Centrifuge at 2000 rpm for 1 min at room temperature (swing-out rotor). This expels the excess buffer and packs down the resin, giving it a white appearance. The height of the resin should come up to *at least* the 0.9 ml mark on the syringe. If the resin level is lower, pipette in more resin suspension, and re-spin as before.
4. Place the spin column in a fresh 15-ml Falcon tube, with a decapped 1.5 microcentrifuge tube at the bottom. (Keep the microcentrifuge cap – you will need this for storing the probe.) Apply the probe sample to the column in a total volume of 50 μl.
5. Centrifuge at 2000 rpm for 1 min at room temperature.
6. Remove the syringe and discard in the radioactive waste. Using forceps, pull out the decapped microcentrifuge tube. The tube should have approximately 50 μl eluate containing the probe.

[a] Modified from Sambrook and Russell (2001), A8.30. volume 3.

[b] Glass balls, 2.5–3.5 mm, BDH, Prod. 332124G. Balls are placed in a bottle and autoclaved. Sambrook and Russell (2001) recommend siliconized glasswool to plug the columns/syringes, but glasswool is toxic, particularly towards the lungs. Glass beads are a much better alternative – easy and safe. Two or three beads will plug the column completely.

[c] Sephadex G-25 DNA Grade resin (Amersham Pharmacia Biotech).

[d] TNES buffer is 0.14 M NaCl; 20 mM Tris, pH 7.5; 5 mM EDTA (8.0); 0.1% SDS. (For a 500 ml stock of TNES buffer, mix 14 ml 5 M NaCl; 10 ml 1 M Tris, pH 7.5; 5 ml 0.5 M EDTA (8.0); 5 ml 10% SDS and 466 ml DEPC-treated water; and autoclave.) The SDS stops mould growing.

Sephadex G-25 is equilibrated as follows: Place roughly 20 g Sephadex G-25 resin in a baked glass bottle. Add approx 100 ml TNES buffer. The exact ratio of resin and buffer is not important. Autoclave. Pour off the excess TNES supernatant. Store the equilibrated resin at 4°C.

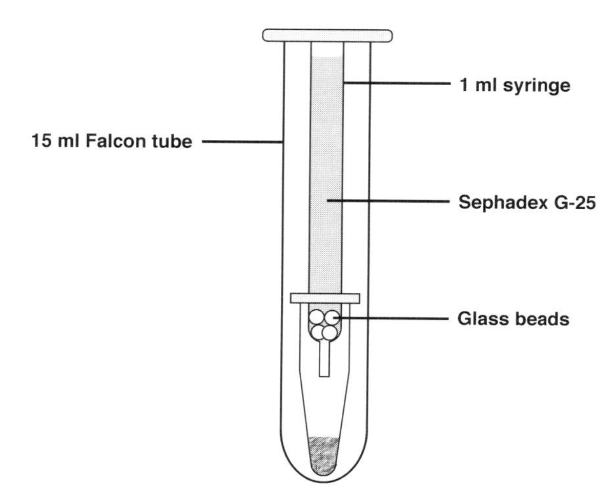

FIG. 1.7. Spin column. An Eppendorf tube, with cap removed, is placed into a 15-ml Falcon tube, followed by a 1-ml disposable syringe. The flanges of the syringe support it in the Falcon tube. A few glass beads are added to plug the column. A suspension of swollen Sephadex G-25 resin is then pipetted into the column until it is full. See Protocol 1.4.1. (Modified from Sambrook and Russell, 2001.)

1.7.3 Removal of Unincorporated Nucleotides: Spin Columns

To remove unincorporated nucleotides from the labelling reaction, we use size-exclusion chromatography, in other words spin columns (Fig. 1.7). Probe and free nucleotides enter the resin at the top of the column. The resin is Sephadex, a bead-formed gel pre-equilibrated with buffer. The resin contains particles with holes that are the doorways to tortuous diffusion paths. Under the pressure of centrifugal force, nucleotides and other small molecules enter these holes, whereas the oligonucleotides are too big to go into the holes and pass directly through the spaces between the resin particles. During a typical 1- or 2-min spin of a 1-ml column, the small molecules stay in the particles at the top of the column, lost in the holes, and the oligonucleotide probe collects in the eluate.

Commercial spin columns can be used, but be careful – not all of these columns work well. At one stage, in both our laboratories the labelling reactions stopped working – there were few counts in the eluate (e.g. 25 000 dpm/μl). This was not due to the isotope, the enzyme or the labelling time, and when we changed the supplier of the spin columns, the labelling reactions were successful again. It seems that the columns had been absorbing the probe. We now use either homemade Sephadex G-25 spin columns (Protocol 1.4.1 and Fig. 1.7), or the Qiagen "Qiaquick Nucleotide Removal Kit". The homemade columns (Sambrook and Russell, 2001, volume 3, appendix 8) work excellently, take only a few minutes to prepare and are cheap (see Protocol 1.4.1). The Qiagen "Qiaquick Nucleotide Removal Kits" are relatively inexpensive and are used according to the manufacturer's instructions.

1.7.4 Simultaneous Labelling of Probes

A considerable advantage of using oligonucleotides compared with other probe types is that many probes can be labelled simultaneously; the number of operator steps is less than those required to label a cRNA probe. For example, if 10 oligonucleotides are to be labelled, it is most efficient to prepare a reaction cocktail (master mix) containing reaction buffer, water, isotope and enzyme in order to minimize pipetting steps. The reaction cocktail should actually be prepared for 11 probes – this allows for small errors in pipetting volumes, and ensures that after the master cocktail is mixed, there is actually enough volume to label 10 probes. Specifically, to have enough buffer for 10 individual 10-μl reactions, 6.6μl $CoCl_2$, 22μl of 5 × reaction buffer, 55 μl sterile water, 16.5 μl [$\alpha^{35}S$]dATP and 11μl of TdT should be mixed in an Eppendorf tube on ice. This is simply the volume of each component in the reaction recipe given in Protocol 1.4 multiplied by 11. It is mixed by gentle vortexing and then centrifuged (brief pulse) to collect all the liquid to the bottom of the tube and 10 μl of the cocktail pipetted successively into 1-μl aliquots of each oligonucleotide to start the reactions.

1.7.5 Storage of Probes

After the probe has gone through the spin column, dithiothreitol (DTT) is added to stabilize ^{35}S probes (Protocol 1.4). Sulfur atoms easily form disulfide bonds with each other; DTT is a reducing agent that prevents this. The half-life of ^{35}S is approximately three months and probes can be stored at –20°C and freeze–thawed whenever they are needed. They can be used for at least up to one month after labelling and possibly longer. The half-life of ^{33}P (see Section 1.7.7) is 29 days and probes can be similarly freeze–thawed and reused over this period. ^{33}P probes do not require DTT to stabilize them.

1.7.6 Troubleshooting the TdT-labelling Reaction

This can be frustrating. Optimal counts after the spin column for ^{35}S-labelled probes are between 100 000 dpm/μl and 300 000 dpm/μl from the 50-μl eluate. Although the TdT reaction usually works, sometimes it fails or works suboptimally (counts at or below 50 000 dpm/μl), or even too well (counts greater than 300 000 dpm/μl). This variability often depends on the batch of the isotope, which may contain impurities that inhibit the enzyme. ^{35}S isotope reagents stabilized with DTT concentrations greater than 1mM are not used, as DTT forms an insoluble complex with cacodylate in the reaction buffer and stops the enzyme activity.

Nucleotides stabilized with Redivue (Amersham) or Easy Tides/Sapphire Blue (NEN) are not used for this purpose; these are marked with a stabilizer dye for keeping the isotope at 4°C; but they can sometimes inhibit TdT. "*In situ* hybridization grade" ^{35}S (stored frozen at –20°C and stabilized with no more than 1 mM DTT) should be used ([α^{35}S]dATP, 1250 Ci/mmol, NEG034H, NEN; [α^{33}P]dATP, 3000 Ci/mmol, NEG312H, NEN).

For ^{35}S, if the counts are below 50 000 dpm/μl, the probe-specific activity will be too low to detect many mRNAs, but it will still work for actin, tubulin and other abundant mRNAs such as MAP-2. If things are not working, the reaction can be left to incubate for a longer time, e.g. 30 min, or the amount of TdT enzyme increased (e.g. by 30%) at the expense of the equivalent volume of water; alternatively, the amount of [α^{35}S]dATP can be increased (e.g. by 30%) at the expense of the equivalent volume of water. Sometimes commercial spin columns absorb some of the probe, giving reduced counts (see Section 1.7.3); in this case homemade Sephadex G-25 columns can be used (Protocol 1.4.1).

In contrast, sometimes the reaction can work too well: probes with counts in excess of 350 000 dpm/μl to 400 000 dpm/μl may give non-specific binding, particularly to granule cells of the cerebellum and other dense cell areas such as the hippocampus. This is probably because the poly[^{35}S]dA tail is too long. In this case a shorter reaction time can be tried or only 1 μl of the [α^{35}S]dATP instead of 1.5 μl (see Protocol 1.4). However, if the cerebellum is not required, such a probe may not be a problem and an autoradiographic signal will be obtained quickly on both X-ray film and emulsion (although of course it is important to make sure that control hybridization is performed in parallel on adjacent sections).

There is considerable variation in enzyme activities depending on the supplier. For example, the Boehringer-Mannheim enzyme requires only 5 min to get probes of the required specific activity. Other sources of enzyme may require longer incubation periods (e.g. 30–60 min at 37°C) to achieve the same result.

1.7.7 ^{33}P-labelled Probes

In principle, α^{33}P speeds up the ISH process when detecting rare mRNAs. The E_{max} energy of emitted β-particles from ^{33}P is 0.25 MeV, whereas the β-particle E_{max} from ^{35}S is 0.17 MeV (Sambrook and Russell, 2001). Thus, exposures with ^{33}P to X-ray film are roughly 1.5 times faster than those for ^{35}S for the same specific activity, but the resolution on X-ray film is comparable. For rare mRNAs, this may give ^{33}P a marginal edge. However, ^{33}P does not work quite as well as ^{35}S on emulsion. We still mainly use ^{35}S.

Figure 1.8 illustrates relative signal intensities of a probe tailed with poly([^{35}S]dA) or poly([^{33}P]dA) using a 30 : 1 molar ratio of dATP to oligonucleotide. The 45mer probe hybridizes to an mRNA encoding the $5HT_{5B}$ receptor, found predominantly in the CA1 sector of the rat hippocampus (Wisden *et al.*, 1993). The ^{35}S exposure (panel A) was four weeks on XAR-5 X-ray film. Figure 1.8B illustrates the same oligonucleotide labelled with ^{33}P, with an exposure time of only 4 days. After four weeks on film, a saturated image results for the ^{33}P-labelled probe (Fig. 1.8C).

1.7.8 ^{32}P-labelled Probes

^{32}P probes give fast results on X-ray film if using a scintillation screen and exposing at –70°C. However, the quality of the autoradiographs is poorer (more out of focus) than with ^{35}S or ^{33}P (e.g. Wisden *et al.*, 1988). β-electrons from ^{32}P are very energetic and move a relatively long way from their source atom – most of the electrons pass through the film and strike the scintillation screen. If this happens at very cold temperatures, it results in light emission, and it is the photons that then expose the X-ray film. The light is even more scattered than the original β-particles – hence the out-of-focus look on the X-ray films. At room temperature, scintillation screens do not work (see appendix 9 of Sambrook and Russell, 2001, for explanation of scintillation screens). So ^{32}P is less sensitive than ^{35}S on X-ray film when exposed at room temperature. It is also less sensitive for sections dipped in emulsions. For those ^{32}P electrons that become trapped in emulsions, the resulting silver grains are more scattered compared with those produced with ^{35}S, and thus resolution over cells is poorer (see Chapter 9, Section 9.5.3, Isotopes and emulsions).

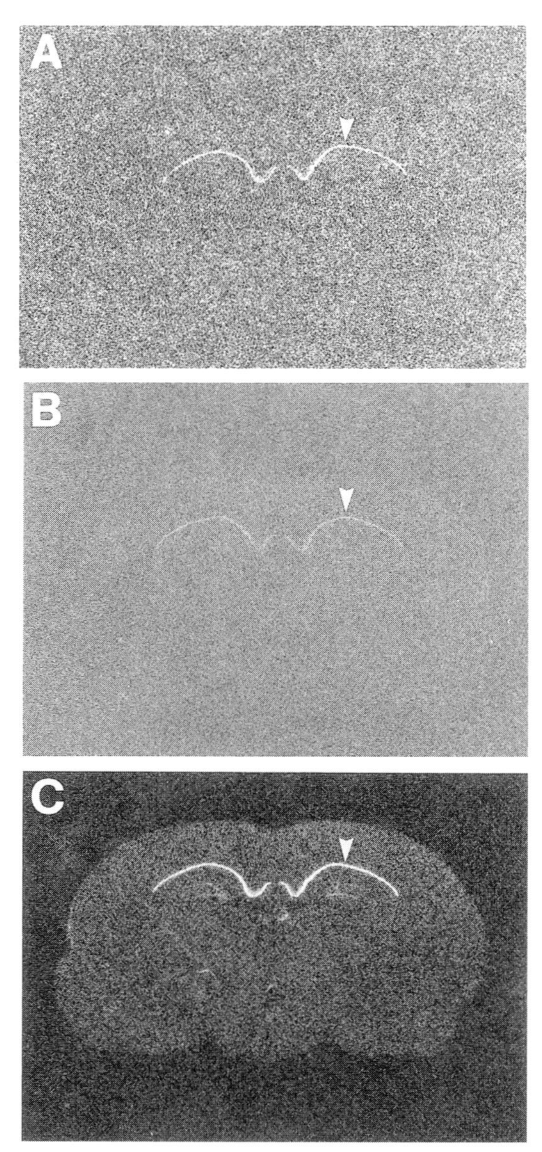

FIG. 1.8. X-ray film autoradiographs (coronal sections, adult rat) comparing results from a poly[^{35}S]dA-labelled oligonucleotide (A) and the same oligonucleotide tailed with poly[^{33}P]dA) (B and C). The oligonucleotide hybridizes to the $5HT_{5B}$ receptor mRNA. In the rat, the gene is mainly expressed in the CA1 pyramidal cells (arrowheads) (Wisden *et al.*, 1993). The longer exposure with ^{33}P allows weaker expression in the dentate granule cells to be seen. See text, Section 1.7.8.

1.7.9 Tailing Oligonucleotides with Digoxigenin-tagged Nucleotides

Tailing oligonucleotides with digoxigenin-tagged nucleotides would be an excellent way of doing non-radioactive ISH. We have tried this, and had no success with it. The tailing reaction works; however, because of the reduced sensitivity, a much higher concentration of oligonucleotide has to be used on the section, and this gives a high background. For non-radioactive oligonucleotides, it is best to use the alkaline phosphatase method described in Chapter 10.

1.8 Hybridization and Washing.

1.8.1 Basic Protocol

The probe is diluted in the hybridization buffer to a concentration of 0.3 pmol/5000 µl (i.e. for a 45mer, 1 pg/µl). This means 1 µl of the probe (from the 50-µl spin column eluate) per 100 µl of hybridization buffer. For a standard 25 mm × 75 mm microscope slide, 100 µl of hybridization buffer is needed after it has been spread out under a coverslip. Thus, 10 microscope slides require 1000 µl of hybridization buffer and 10 µl of probe.The hybridization buffer is prepared in 50-ml batches in 50-ml sterile polypropylene Falcon tubes (Protocol 1.7). This amount of buffer is enough for 500 standard slides. Hybridization buffer is stored at 4°C, and is stable for long periods (at least six months). With prolonged storage at 4°C, various components of the buffer may precipitate out as a "white smear". This is not a cause for concern, but it is a good idea always to shake/vortex vigorously before use.

For the hybridization, strips of parafilm (cut to about 20 mm × 55 mm) are used as coverslips to spread out the hybridization liquid (Protocol 1.5 and Fig. 1.9). Not all sealing films are suitable; some melt at the hybridization temperatures. The hybridization chamber (usually a sealed plastic Petri dish) is kept moist using a ball of tissue paper saturated in 50% formamide/4 × SSC (see Fig. 1.10).

After hybridization, and before immersing slides in wash solutions, the parafilm coverslips are pulled gently off the section with blunt-ended forceps, and the slides are then transferred directly into room temperature 1 × SSC solution (Protocol 1.6). It is important not to allow the sections to dry between the removal of the coverslip and the immersion

PROTOCOL 1.5 APPLICATION AND HYBRIDIZATION OF PROBES TO SECTIONS

1. Remove the sections from 95% ethanol storage and allow to air-dry for anywhere between 15 min and half an hour. The exact time is not critical. Label the slides with the appropriate probe designation using a pencil on the frosted end of the slide. These pencil marks do not come off during any of the subsequent steps. If using sections of retina or adrenal gland, you may wish to pretreat the slides with acetic anhydride and/or chloroform prior to hybridization. If this is the case, transfer the sections straight from the 95% ethanol storage into 1 × PBS and go to step 2, Protocol 1.10.
2. In an Eppendorf tube (for smaller volumes) or a 15-ml Falcon tube (for larger volumes), dilute the radiolabelled probe in hybridization buffer.[a] Because of the high viscosity of the hybridization buffer, the probe solution has a tendency to immediately float to the surface after addition to the hybridization buffer mix. Thus, it is very important to vortex the probe/hybridization buffer vigorously to achieve an even mix.
3. Apply 100 μl of probe/hybridization buffer to each slide.[b] If several sections are on each slide, make sure that there is some hybridization buffer on all the sections before putting on the parafilm.
4. Gently lower a parafilm coverslip over the drop of hybridization buffer. The liquid should spread smoothly under the coverslip. Remove any large air bubbles by very gentle pressing with blunt-ended forceps, but any remaining small air bubbles are not a problem as they tend to disappear during the hybridization step.[c]
5. To maintain humidity, saturate a small piece of tissue/filter paper with 50% formamide/4 × SSC[d] and place this in the Petri dish[c] or Nunc Bio-Assay dish[c] as well. Place the lid tightly on the dish and incubate overnight at 42°C. It is advisable to further seal the lid of the dish by stretching layers of parafilm around the lid.

[a] Probes are used at a low concentration in the hybridization buffer. Dilute 1 μl (or 2 μl) of probe from the 50 μl spin column eluate into 100 μl hybridization buffer. Use 100 μl hybridization buffer/slide. Also add 1 μl 1 M DTT/100 μl hybridization buffer. It is best to pre-aliquot the 1 M DTT stock, rather than keep using one stock tube, as DTT goes off with multiple freeze–thawing – this causes an increased background on the sections (U. Amtman, personal communication). If the background signal with ^{35}S-labelled probes is any higher than barely perceptible, it can frequently be reduced by increasing the amount of DTT to 4 μl 1 M DTT/100 μl hybridization buffer.
Preparation of competition controls. Dilute the radiolabelled probe exactly as in footnote *a* but also add 2 μl of the corresponding unlabelled probe directly from the 0.3 pmol/μl stock for every 100 μl of hybridization buffer to generate a 100-fold excess of unlabelled probe.

[b] Use Parafilm "M" laboratory film. American National Can TM., Greenwich, CT 06836, USA.

[c] Slides are laid horizontally in disposable transparent plastic Petri dishes (for small numbers of slides) or Nunc Bio-Assay dishes (size: 243 × 243 × 18 mm) (if using a large number of slides). If these dishes are placed on a black/dark surface it becomes easy to see any air bubbles. Use parafilm strips cut to roughly 20 mm × 55 mm as

coverslips. It is best to precut these prior to the experiment. Before use, remember to peel off the paper backing of the unexposed (virgin) side of the parafilm. Use this side to contact the hybridization liquid.

[d] We use 50% formamide/4 × SSC ("tissue wet") to humidify the hybridization chamber (same osmolarity as the hybridization buffer, to prevent any distillation between the two solutions). This "tissue wet" solution can be stored at room temperature, and for the purpose of humidification, does not need to be made from DEPC-treated SSC. (The formamide may not be necessary, and 4 × SSC can be used equally well.)

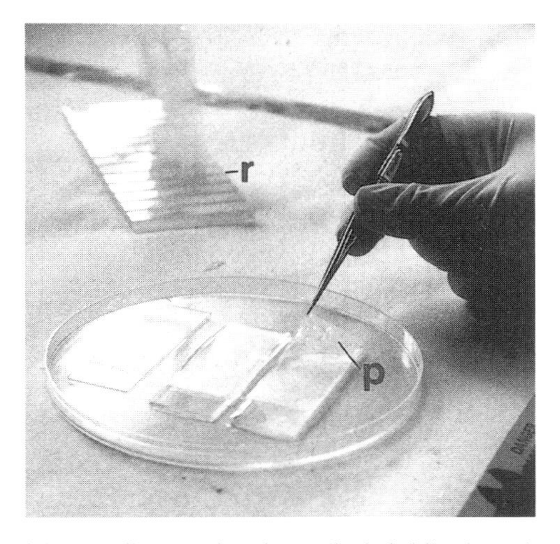

FIG. 1.9. Parafilm (p) coverslips are placed over the hybridization mixture on slides. A perspex rack/plate for holding the slides in an upright position is shown in the background (such a rack is also illustrated in the BDH/Merck Laboratory Supplies Catalogue).

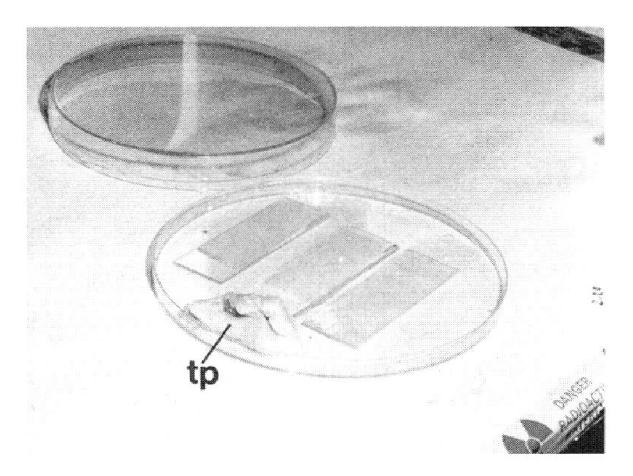

FIG. 1.10. A Petri dish hybridization chamber. A tissue paper (tp) saturated with 50% formamide/4 × SSC is in the Petri dish.

Protocol 1.6 Washing Sections after the Hybridization

All procedures can be non-sterile at this point.

1. Gently peel off the parafilm coverslips with blunt-ended forceps. Transfer the slides into 250 ml 1 × SSC at room temperature. This step removes most of the unhybridized probe, prior to the higher stringency wash in step 2.[a]
2. Transfer the rack of slides into 250 ml pre-warmed 1 × SSC at 55°C. It is convenient to have the continental troughs in a 55°C water bath. Leave the sections washing for half an hour. The exact time is not critical. Agitation during washing is not normally required, although if there is any indication of a relatively high background signal, it may be advantageous to use a shaking water bath, particularly when the rack will be full and the slides tightly packed. Addition of DTT is not required. Sections can also be washed with 1 × SSC at 60°C with no appreciable difference in signal.
3. Transfer the rack through a brief (couple of seconds each) series of room temperature rinses in 1 × SSC, 0.1 × SSC,[b] 70% ethanol, 95% ethanol (250 ml of each solution). Allow sections to air-dry (half an hour).
4. Expose sections to X-ray film or dip in emulsion (Protocol 1.9).

[a] Do not allow sections to dry after removal of coverslip (see text, Section 1.8).
[b] 0.1 × SSC step is required to prevent salt precipitation on sections in ethanol.

Protocol 1.7 "Maximalist" Oligonucleotide Hybridization Buffer

The hybridization buffer is 50% formamide, 4 X SSC, 10% dextran sulfate, 5 x Denhardt's, 200μg/ml acid–alkali cleaved salmon sperm DNA, 100 μg/ml long-chain polyadenylic acid, 25 mM sodium phosphate pH 7.0, 1mM sodium pyrophosphate.

In a sterile graduated and screw-capped 50-ml polypropylene tube, add the following:

25 ml 100% formamide[a]
10 ml 20 × SSC[b]
2.5 ml 0.5 M sodium phosphate pH 7.0[c]
0.5 ml 0.1 M sodium pyrophosphate
5 ml 50 × Denhardt's solution[d]
2.5 ml 5 mg/ml acid–alkali hydrolysed salmon sperm DNA[e]
1ml 5 mg/ml polyadenylic acid[f]
5g dextran sulfate[g]
After the dextran sulfate has dissolved, adjust to 50 ml with DEPC-treated water.

The dextran sulfate takes several hours to dissolve completely, with occasional hard vortexing. Between vortexing, place the tube on a rocker platform.

Store the hybridization buffer at 4°C in a sterile polypropylene (Falcon) tube. It keeps for at least six months and probably longer. Shake well before use. Do not boil the hybridization buffer prior to use.

[a] Use Fluka formamide. This brand requires no de-ionization prior to use. Use straight out of the stock bottle (stored at room temperature in light-tight bottle).

[b] 20 × SSC is 3 M NaCl, 0.3 M Na citrate, pH 7.0 with NaOH. Filtered, DEPC-treated and autoclaved. See Sambrook and Russell (2001).

[c] 0.5 M sodium phosphate pH 7.0 is prepared by mixing 0.5 M Na_2HPO_4 and 0.5 M NaH_2PO_4 until the pH reaches 7.0. The solution is filtered, DEPC-treated and autoclaved.

[d] 50 × Denhardt's is 5 g polyvinylpyrrolidine (PVO), 5 g bovine serum albumin (BSA), 5 g Ficoll 400/500 ml of DEPC-treated water. Store in aliquots at –20°C. See Sambrook and Russell (2001).

[e] Acid-cleaved salmon sperm DNA is prepared as follows (protocol from Dr Olof H. Sundin): Transfer 1 g salmon sperm DNA (Sigma, deoxyribonucleic acid sodium salt type III from salmon testis) into a 50-ml Falcon tube. Add 15 ml DEPC-treated water, shake and allow the DNA to soak for 15 min to 2 h at room temperature. Add 2.5 ml 2 M HCl (room temperature). The DNA forms a white precipitate; shake until the precipitate collects into a ball; knead the ball with a pipette for 1 min; leave at room temperature for 25 min; add 5 ml 2 M NaOH and shake to resuspend the DNA, which should become soluble in the alkali; place the tube at 50°C for 15 min to help dissolve the DNA. After this heating step, add 175 ml DEPC-treated water and 20 ml 1 M Tris–HCl, pH 7.4. If necessary titrate with 2 M HCl until the pH is between 7.5 and 7.0. The concentration will be somewhere between 3 and 5 mg/ml. The solution is too viscous to filter. Store the solution at –20°C, and freeze–thaw as required. The exact concentration of hydrolysed DNA in the hybridization buffer mix is not critical: add 2.5 ml of whatever DNA stock solution you have when preparing the hybridization buffer mix.

[f] Dissolve 100 mg of polyadenylic acid (5′), potassium salt (Sigma no. P-9403) in 20 ml of DEPC-treated water, to give 5 mg/ml stock. Store in 1-ml aliquots at –20°C.

[g] Dextran sulfate, sodium salt, molecular biology grade (Amersham Pharmacia Biotech).

---

of the slide into the wash solution; otherwise the probe may "bake" permanently to the section and slide. This causes a terrible background signal (Fig. 1.18E). It is better to remove coverslips individually and transfer the slides one at a time into the washing buffer, or to remove the coverslips once the slides are immersed in the washing buffer. For ^{35}S-labelled probes, most protocols suggest having 10mM DTT or β-mercaptoethanol in the wash solutions to prevent coupling of ^{35}S to the section. However, we find their addition to be of no value and they are routinely omitted.

1.8.2 Increasing the Sensitivity by Hybridizing Multiple ^{35}S-Labelled Oligonucleotide Probes Simultaneously to the Same Section

If the target mRNA is rare, sometimes only a faint autoradiographic signal is produced, even after long exposure to X-ray film (Fig. 1.11). Sensitivity is improved by hybridizing multiple oligonucleotide probes to the same section (e.g. Trivedi *et al.*, 2001). For example, the mouse NPY1 receptor mRNA is not abundant in mouse brain. Figure 1.11 shows the result of individually hybridizing four different ^{35}S-labelled oligonucleotides to the mouse NPY1 receptor mRNA in four parallel sections. Each oligonucleotide recognizes a different part of the mRNA. All four patterns are the same, but faint. There is a signal over dentate granule cells, but none in the cortex. The bottom panel shows the result of adding all four probes to the same hybridization buffer: now clear labelling of the cortex is seen (layers II and V/VI) and the overall image intensity is suitably stronger. In this experiment, 1 μl of each ^{35}S-labelled oligonucleotide (from the 50-μl spin column eluate) was added per 100 μl of hybridization buffer (A. Oberto and C. Eva, University Torino, unpublished). The same principle of using multiple probes applies to alkaline phosphatase-labelled oligonucleotides (see Chapter 10, Section 10.4.1 and Fig. 10.4).

1.8.3 Components of the Hybridization Buffer

The key components of the hybridization buffer are formamide and Na^+ (in the form of SSC and dextran sulfate). Formamide makes it possible to lower the temperature at which hybridization occurs, thus preserving section morphology. High salt concentrations promote hybridization rate, and dextran sulfate also serves to increase rate of hybridization (see Wahl *et al.*, 1987 and Sambrook and Russell, 2001 for reviews). Dextran sulfate works by forming aggregates that exclude water; this forces the probe into higher concentrations in the aqueous part, thus increasing the hybridization rate.

Hybridization stringency is determined by temperature, percentage formamide and Na^+ ion concentration (see Chapter 10, Section 10.3.1). For a given hybridization temperature and length of oligonucleotide, a lower percentage formamide mix corresponds to a lower stringency. In concrete terms, for a 45mer hybridizing to mRNA, 50% formamide/4 × SSC at 42°C is more stringent than 30% formamide/4 × SSC at 42°C. The latter might be more appropriate for a 26mer. For a review of conditions for hybridizing oligonucleotides to nucleic acids immobilized on membranes see Lathe (1985), Albretsen *et al.* (1988), and Sambrook and Russell, 2001, (volume 2, chapter 10). The other components of the hybridization buffer

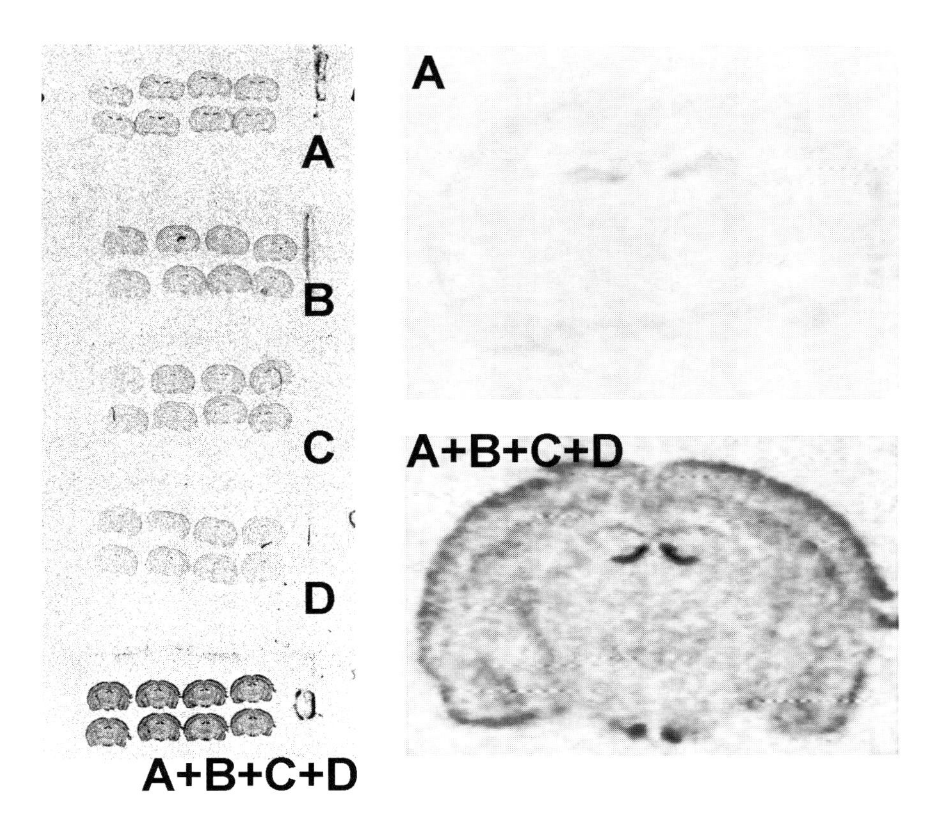

FIG. 1.11. Increasing ISH sensitivity by hybridizing multiple oligonucleotide probes simultaneously. The left-hand panel shows the primary X-ray film autoradiographs resulting from hybridizing individual oligonucleotides (A, B, C and D) to different regions of the mouse NPYl receptor mRNA (Trivedi *et al.*, 2001) in mouse coronal brain sections (multiple sections are on each slide). The signal is strongly increased by hybridizing all four oligonucletides together (A+B+C+D). These all result from the same exposure to one piece of X-ray film. The right-hand pictures are blow-ups of one of the autoradiographs from probe A, and the four probes together. (A. Oberto and C. Eva, University of Torino, Italy, unpublished; see also Trivedi *et al.,* 2001).

cocktail can prevent non-specific hybridization of probe to other RNAs or protein/lipid components in the section. They are at such high concentrations in the buffer that they compete with the probe for every non-specific site. For example, the polyadenylic acid is added to the hybridization buffer to compete with the polyadenylic tail of the probe. Similarly, salmon sperm DNA and Denhardt's solution are nucleic acid-blocking reagents (see Sambrook and Russell, 2001). Hydrolysed salmon

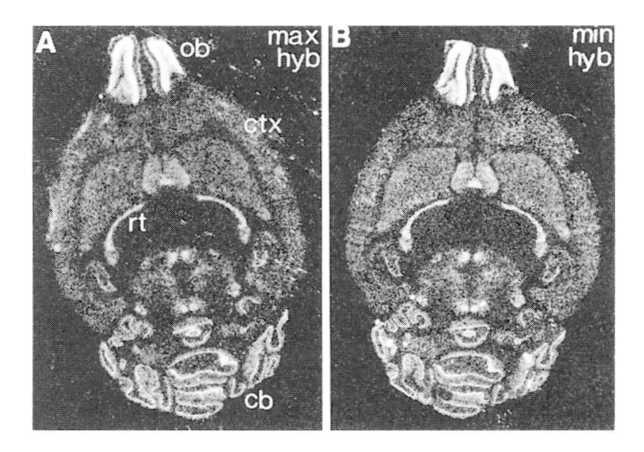

FIG. 1.12. X-ray film autoradiographs (rat brain, horizontal sections) comparing the effects of hybridizing a ^{35}S-labelled oligonucleotide recognizing the mRNA encoding rat glutamic acid decarboxylase (GAD67) in either (A) "maximalist" buffer (Protocol 1.7) or (B) "minimalist" buffer (Protocol 1.8). Exposure time was identical for both images (one week). Cb, cerebellum; ctx, neocortex; ob, olfactory bulb; rt, reticular thalamic nucleus.

sperm DNA (Protocol 1.7, footnote *e*) is used because it is already broken into small pieces that compete more effectively with oligonucleotides. Pyrophosphate competes with any remaining free nucleotides after the spin column, and sodium phosphate pH 7.0 keeps the mix buffered for optimal hybridization. Together, these components add to a collective overkill to prevent non-specific binding in a maximal hybridization buffer (Protocol 1.7).

Many of these extra ingredients are redundant. Figure 1.12 shows the results of an experiment in which an antisense 45mer oligonucleotide designed to hybridize to the rat mRNA encoding glutamic acid decarboxylase (GAD67) was hybridized on parallel horizontal rat brain sections in two different hybridization buffers. The minimalist buffer contained only 50% formamide/4 × SSC/10% dextran sulfate (Fig. 1.12B). The other buffer was the traditional maximal hybridization buffer (Protocol 1.7; Fig. 1.12A). Thus, at the low probe concentrations used in our ISH procedure, the minimalist buffer is as effective as the traditional one, at least for abundant mRNAs. The minimalist hybridization buffer recipe is given in Protocol 1.8. We use this buffer routinely, simply because it is easier to make. On certain tissues, such as retina, adrenal gland, embryo and spinal cord, the maximalist buffer gives better results. For detecting relatively rare mRNAs, a compromise buffer (minimalist buffer with the addition of polyadenylic acid and sodium pyrophosphate, as described for the maximalist buffer) ensures a low background signal after long exposure times.

PROTOCOL 1.8 "MINIMALIST" HYBRIDIZATION BUFFER

The hybridization buffer is 50% formamide/4 × SSC/10% dextran sulfate.[a] See also Protocol 1.7 for exact specifications of reagents.
To a sterile graduated and screw-capped 50-ml polypropylene tube, add 25 ml 100% formamide (Fluka) and 10 ml 20 × SSC. Add H_2O to approx. 40 ml. Add 5 g dextran sulfate. After dissolution (takes several hours with occasional vortexing, place on a rocker platform when not vortexing), adjust volume to 50 ml.
Store at 4°C. Buffer is stable for at least six months and probably longer.

[a] For the detection of rare mRNAs in brain tissue, where it can be important to keep non-specific hybridization to an absolute minimum even after long exposure times, one of us (BJM) prefers an "intermediate" hybridization buffer, consisting of the "maximalist" hybridization buffer from Protocol 1.7, without the Denhardt's solution and the salmon sperm DNA (final volume adjusted to 50 ml as usual).

1.8.4 HYBRIDIZATION AND POST-HYBRIDIZATION WASHING CONDITIONS

The kinetics of how oligonucleotides hybridize to membrane-bound nucleic acids has been well studied (Lathe, 1985; Albretsen *et al.*, 1988). The length of the oligonucleotide and its nucleotide composition are key variables influencing the optimal hybridization conditions. However, for ISH, the hybridization kinetics are probably quite different from those occurring on membranes, and so the predictive utility of these rules is of less value. An empirical approach works best. For example, when using 36mer oligonucleotides to distinguish between closely related splice forms of glutamate receptor mRNAs (Sommer *et al.*, 1990), two of these probes (D Flop and A Flop) differed by only 3 out of 36 nucleotides. Yet these probes give different distributions in rat brain (Sommer *et al.*, 1990). So clearly the hybridization conditions used in this experiment (50% formamide/ 4 × SSC/10% dextran sulfate at 42°C), combined with a low probe concentration are probably enough to achieve a distinction even with a three nucleotide difference. Conversely, our conditions make it difficult to detect mRNAs when cross-hybridizing between species, since a few dispersed nucleotide mismatches will reduce efficiency of hybridization. Furthermore, for the 36mers, it essentially makes no difference to the qualitative autoradiographic pattern whether the sections are washed at 0.1 × SSC at 60°C or 1 × SSC at 60°C, except that sections washed at higher stringency produce weaker signals. It is the stringency of the hybridization, and not the washing conditions, that is critical in determining specificity.

1.9 Exposure to X-ray Film and Dipping in Emulsion

After washing and dehydration (Protocol 1.6), slides are allowed to air-dry (e.g. 30 min to 1 h) and are then ready to be exposed in a standard X-ray film cassette or dipped in liquid photographic emulsion (see Section 1.9.4 and Protocol 1.9).

1.9.1 The Usefulness of X-ray Film Autoradiographs: Interpreting X-ray Film Patterns

For vertebrate brains, X-ray film resolution should be the initial choice to obtain a general picture of the regions where a particular gene is expressed, to compare the distribution of different mRNAs, and for ease of presentation (see Section 1.12). For some regions of vertebrate brains and embryos (see, for example, Figs 3.2 and 3.4), X-ray film alone is enough to determine which type or layer of cells is expressing the gene. Many genes show highly distinctive expression patterns, localized to just a few brain areas (e.g. the distribution of proenkephalin mRNA in Fig. 1.13A). Some genes are expressed with a bias towards the telencephalon (e.g. Fig. 1.13B); others toward the hindbrain. Spotty hybridization signals may indicate mRNAs concentrated in GABAergic interneurons (e.g. somatostatin, NPY, glutamic acid decarboxylase) (Fig. 1.12 and 1.13C). Ubiquitously expressed mRNAs (e.g. many cytoskeletal or housekeeping genes) show an even distribution that mirrors cell density (Fig. 1.13E); some mRNAs show a particularly high expression in oligodendrocytes and related cells, and hence give a strong specific signal in the main fibre tracts of the brain (Fig. 1.13G). X-ray film resolution is good enough to show if mRNA is present in dendrites (Fig. 1.14), for example in the hippocampus (Garner *et al.,* 1988; Johnston and Morris, 1994; Herb *et al.,* 1997; Roberts *et al.,* 1998; Paterlini *et al.,* 2000). Even for sections as small as the spinal cord, X-ray film analysis provides useful information (e.g. Tölle *et al.,* 1993; see Chapter 5). X-ray film images are easily quantifiable (see Chapter 9, Section 9.4).

1.9.2 X-ray Film Types

Autoradiography films are made by coating clear plastic sheets with photographic emulsion. β-particles from the sections enter the emulsion and cause latent development of silver halide crystals. In earlier studies we used Kodak XAR-5 film. This film, coated with emulsion on both sides, has a grey

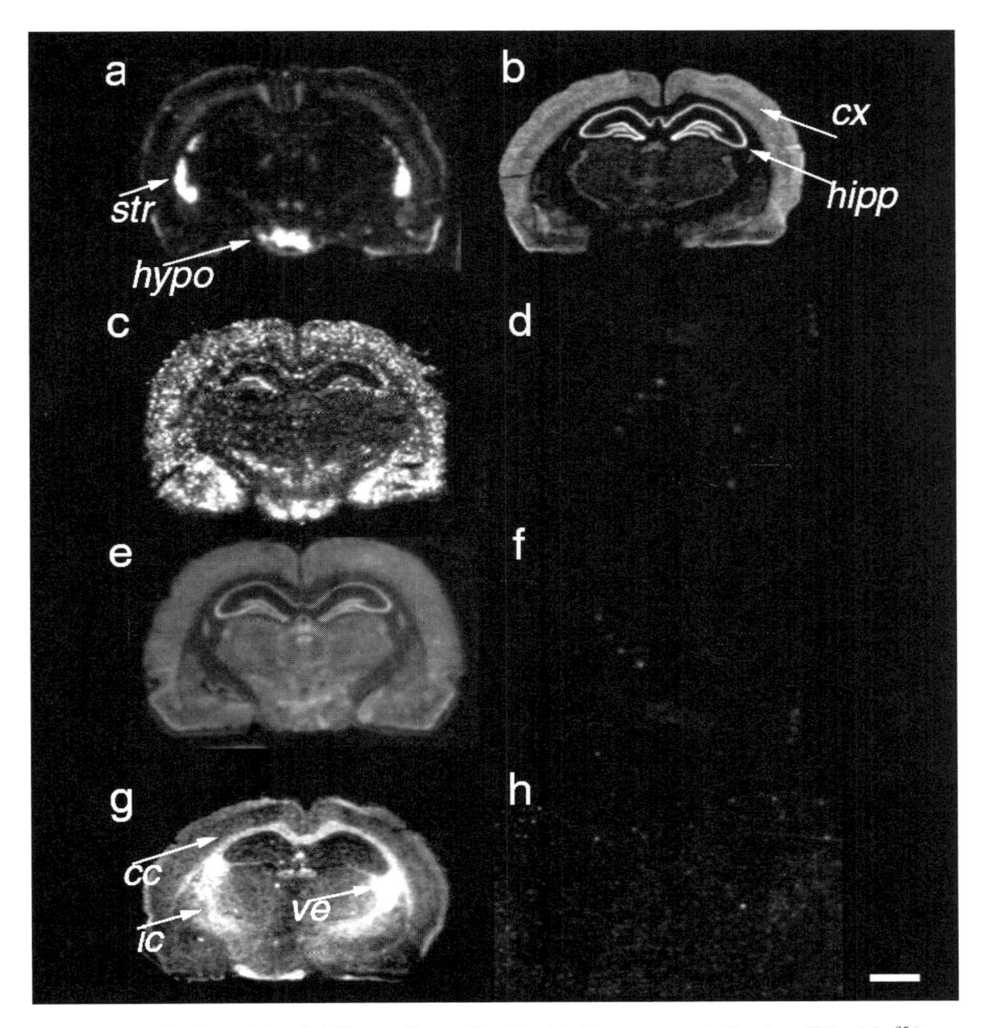

FIG. 1.13. Examples of different X-ray film distribution patterns following ISH with ^{35}S-labelled oligonucleotide probes. Coronal sections of rat brain are shown. (A) Proenkephalin mRNA has a distinctive and highly heterogeneous distribution in the brain; proenkephalin mRNA is expressed at high levels in the striatum (str) and hypothalamus (hypo) (see Morris *et al.*, 1988). (B) An mRNA encoding a post-synaptic density protein with a widespread, but not entirely uniform, distribution in the brain. The gene is expressed at higher levels in the cerebral cortex (cx) and hippocampal formation (hipp). (C) Somatostatin mRNA has a heterogeneous distribution concentrated in spot-like areas – a characteristic of mRNAs expressed at high levels in subpopulations of inhibitory interneurons. (D) Adjacent section to C, hybridized with the labelled somatostatin probe and an excess of unlabelled probe (see Section 1.10). The absence of signal suggests that the signal in C results from specific hybridization. (E) α-Tubulin mRNA has a uniform distribution in the brain. (F) Competition control, as in D, suggesting that the signal in E results from specific hybridization. (G) An mRNA expressed primarily in white matter tracts, such as the corpus callosum (cc) and internal capsule (ic), and also in ventricular ependymal cells (ve). (H) Competition control, as in D,

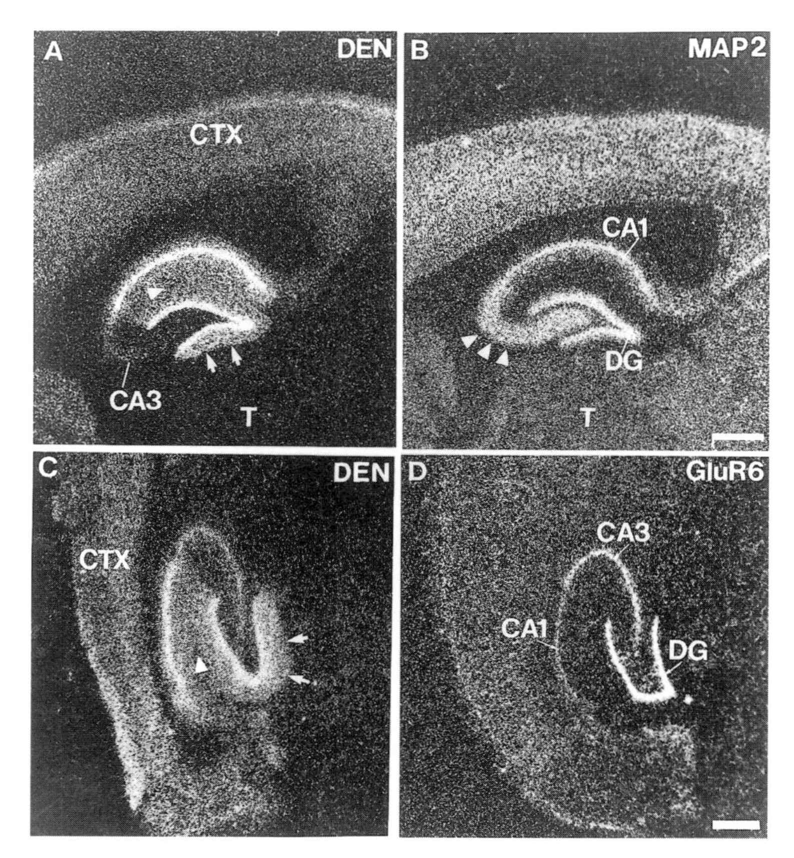

FIG. 1.14. Dendritic mRNAs in the rat hippocampus. (A,B) Sagittal plane; (C,D) horizontal plane. In A and C, the arrowhead marks the distribution of silver grains over the stratum lucidum and stratum radians, indicating dendritic labelling. In B, the distribution of MAP-2 differs markedly from dendrin. MAP-2 transcripts are not found in the very distal portions of CA1 pyramidal dendrites in the stratum radians or in distal dendritic portions of the dentate granule cells. However, in contrast to dendrin, the MAP-2-expressing CA3 pyramidal cells have MAP-2 mRNA in their dendrites, as suggested by the broad hybridization band in this zone. (D) The hippocampal distribution of the glutamate receptor subunit GluR6 mRNA, a nondendritically located mRNA, is shown for comparison. CTX, neocortex; DG, dentate granule cells; T, thalamus. All images are X-ray film autoradiographs printed onto photographic paper. (Original descriptions in Garner *et al.*, 1988; Wisden and Seeburg, 1993; Herb *et al.*, 1997. Picture reproduced from Herb *et al.*, 1997.)

FIG. 1.13 *continued*

suggesting that the signal in G results from specific hybridization. Note that white-matter distributions similar to that shown in G can sometimes be due to non-specific binding, but in such cases they are not displaced in the competition control (see Fig 1.18F). These images were digitally scanned and assembled in Adobe Photoshop (see Section 1.12.2). Scale bar 2 mm.

FIG. 1.15. A dipping chamber for coating slides with photographic emulsion. A standard microscope slide (with sections) is shown for scale. One end of the slide (marked by arrow) is frosted, making it useful for labelling with pencil.

and slightly grainy base, and can be developed in an automatic Xomat machine. It is still the main film used for ^{32}P exposures in most molecular biology labs. However, we now use BioMax MR ^{35}S-autoradiography film (also Kodak) for ISH. Better images can be obtained with BioMax MR. BioMax MR film has a clear base that increases the contrast between the developed image and the film base. This film is also more sensitive (two times) than XAR-5. Unlike XAR-5 film, BioMax film has flat silver halide crystals that cover more surface area; this means the crystal has an increased chance of capturing a β-particle, thus increasing sensitivity. The emulsion on BioMax MR film is only on one side. This is the matt surface, and it is this matt surface that should be placed facing the sections. As for XAR-5, BioMax MR film can be developed with a machine. Alternatively, Hyperfilm βmax film (Amersham Pharmacia Biotech) can be used; this single-sided high-resolution film, designed for ^{35}S and ^{33}P, must be developed manually (see also Chapter 9, Section 9.4.2, Resolution of film images).

1.9.3. How to Expose X-ray Films to Slides; Exposure Times

For X-ray film exposure, slides are first stuck to a cardboard sheet (convenient sheets are often supplied as stiffeners in boxes of X-ray film). The slides are attached along the base using autoclave marker tape or a similar tape. It should be a tape that can easily be peeled off again if the slides are to be dipped in emulsion later. If using single-sided film, such as BioMax MR

(Section 1.9.2), the matt surface must be placed against the slides, as it is this matt surface that has the emulsion (in other words, "shiny side up" – as seen by the reflection of the darkroom safe-light). The cassette should be tight-fitting and must press the film and sections firmly together, otherwise "out-of-focus" images may result. If this occurs, an extra sheet of cardboard can be added. Scintillation screens in the cassette are useful for making firm packing, but they do not increase the sensitivity (see below). The cassettes are exposed at room temperature in a vibration-free environment. Mechanical impacts (bangs and jolts) to the cassette may cause the film to move slightly relative to the slides and a double image will result. If such an accident happens, it is best to develop the film and re-expose. Note: the β-electrons from ^{35}S and ^{33}P are completely absorbed by the film, so exposure at –70°C with scintillation screens is not required, and is in fact futile.

Exposure time depends on both mRNA abundance and probe specific activity. It is subjective for how long you want to "burn in" the image (see also Section 1.12.3). For ^{35}S-labelled probes, exposures vary anywhere between overnight (for abundant mRNAs such as actin or MAP-2, or certain neuropeptide mRNAs, e.g. proenkephalin mRNA) six weeks and (for rare mRNAs, e.g. nicotinic receptor subunit mRNAs, growth factor mRNAs) to get a good signal. A preliminary exposure of one week on X-ray film is done first. More often than not, this results in an image suitable for printing onto photographic paper or for electronic scanning (see Section 1.12). ^{33}P-labelled probes will require, on average, one-third shorter exposure times (see Section 1.7.8). As described in Section 1.8.2, hybridizing multiple oligonucleotide probes together can reduce exposure times.

1.9.4 Cellular Resolution Using Photographic Emulsion

To achieve cellular resolution with radioactive ISH methods, which is sometimes essential, sections have to be coated (dipped) in photographic emulsion (Figs 1.15 and 1.16 and Chapter 9, Section 9.5.5) (see also Figs 4.2, 5.3, 6.3, 9.2IV, 9.3, 10.4E,F, 10.9 and 10.10). For example, on the basis of X-ray film autoradiographs alone, it is not possible to determine if the $GABA_A$ receptor γ1 subunit mRNA is in Purkinje cells or Bergmann glial cells, since both cell body types are co-localized in the cerebellum (Fig. 1.16). Dipping is a classical autoradiographic technique (Rogers, 1979). The slide-mounted sections are literally dipped into liquid photographic emulsion, the same type of emulsion that is used to coat the plastic sheets which make X-ray film; during exposure, β-electrons from radioactive atoms enter the emulsion and cause latent development of the silver atoms. After chemical development and fixation, a thin film of transparent emulsion is left

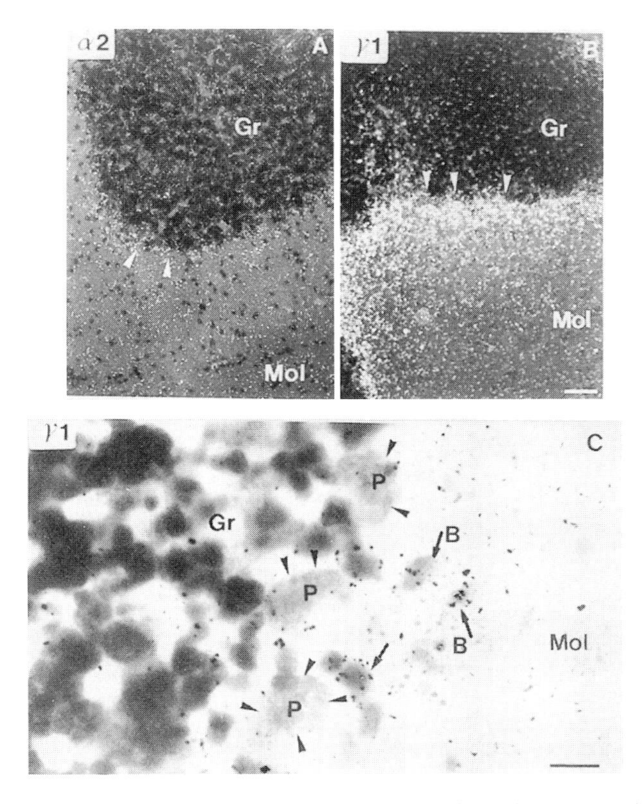

FIG. 1.16. Detection of mRNA encoding the γ1 subunit of $GABA_A$ receptors in Bergmann glia of the rat cerebellum using emulsion autoradiography. The α2 subunit and γ1 subunit probes hybridize only to the Purkinje cell layer as assessed from X-ray film autoradiographs (see Laurie *et al.*, 1992a). Dipping the slides into photographic emulsion confirms that the signal results from hybridization to the Purkinje cell layer. Under dark-field optics, the α2 signal is a halo of silver grains along the border between the granule and molecular cell layers; the granule cells are unlabelled (A). For the γ1 subunit mRNA, a dense cluster of silver grains originates at the granule cell/molecular layer border and extends out into the molecular layer (B). High-power bright-field optics using a 100× lens with immersion oil shows the Purkinje cells (P, black arrowheads) are unlabelled by the γ1 probe (C). Silver grains cluster over small cells surrounding the Purkinje cells, with the granule cells unlabelled. The position of these labelled cells suggests that they are Bergmann glia. Pictures were obtained with a Zeiss Axioplan microscope. (Reproduced with permission from Laurie *et al.*, 1992a. ©1992 by the Society for Neuroscience.)

covering the section – under the microscope you can see silver grains clustered over the sites where the probe has bound to the section. After development, the slide-mounted sections are usually stained with thionin, or some equivalent Nissl stain. If the experiment has worked, the silver grains are clustered over counterstained cell bodies. The silver grains can be viewed under normal light (bright-field) (e.g. Figs 1.16C, 7.4, 9.3 and 10.4E,F) or under reflected light (dark-field) (e.g. Figs 1.16A,B and 6.3C and F). The bright-field conditions are best for looking at cells under very high-power magnification (e.g. Figs 4.2, 6.3 and 9.3) – at low power it can be difficult to photograph the silver grains. Dark-field illumination, in which the silver grains shine white, is best for an overall view (e.g. Figs 1.16A, 5.4, 6.3 and 9.2IV). The combined protocol for dipping hybridized sections in photographic emulsion, and subsequent exposing, developing and thionin staining is given in Protocol 1.9.

Quantification

Changes in mRNA levels can be quantified by counting silver grains, although eliminating false-positive or false-negative results requires care (see Chapter 9, Section 9.5). The autoradiographic signal is as robust and reproducible as X-ray film images, and is equally applicable to semiquantitative studies (Morris, 1997; Simpson and Morris, 2000).

Sensitivity

Coating slides with photographic emulsion is less sensitive than exposure to X-ray film. With ^{35}S-labelled probes, a "rule of thumb" is that slides dipped in emulsion require five times longer than exposure of the same section to X-ray film. Specifically, a one-week exposure to X-ray film needs five weeks with emulsion. In practice we usually expose dipped sections for a minimum of 8–12 weeks to generate strong clustering of silver grains over cell bodies. This is a method designed to strengthen your patience!

Double-labelling

Probes tagged with alkaline phosphatase or digoxigenin eliminate the need to use emulsion for qualitative studies, and the long delays before seeing the result. Non-radioactive ISH only works for abundant targets, but it does have the advantage of being able to localize two mRNA types in the same cell. To do this, emulsion autoradiography and non-radioactive methods are combined by co-hybridizing ^{35}S-labelled and alkaline phosphatase-labelled oligonucleotides (see Chapter 10, Section 10.7 and Figs 10.9 and 10.10). The least abundant mRNA should be detected with ^{35}S.

PROTOCOL 1.9 AUTORADIOGRAPHY OF SECTIONS USING PHOTOGRAPHIC EMULSION

1. The emulsion is mixed on the day of use. All procedures are performed under safe-lighting using Kodak 6B or equivalent filters. Pre-warm 25 ml water/0.5% glycerol in a 50-ml polypropylene screw-top Falcon tube to 43°C. The entire procedure uses a water bath.
2. Add solid emulsion shreds[a] to the pre-warmed water/glycerol so that it displaces the liquid to the 50 ml mark. This makes a 1 : 1 ratio of emulsion water.
3. Wrap the Falcon tube in aluminium foil and allow the emulsion to melt for half an hour at 43°C.
4. *Gently* invert the mixture several times in order to produce a homogeneous solution. Avoid creating bubbles.
5. Filter mixture through muslin cloth into the dipping chamber[b] and allow to stand for several minutes to remove air bubbles. The dipping chamber is also maintained at 43°C.
6. Dip slides individually into the chamber (Fig. 1.15), allow to drain and then place to dry in a rack over a humid environment (damp tissue paper). Slides are allowed to dry in complete darkness for 2–3 h.
7. Transfer slides to light-tight slide boxes, e.g. BDH/Merck "staining rack" and "troughs". These are black polyacetate boxes with push-fit lids (size 100 × 85 × 55 mm) each containing a sachet of silica gel. Seal boxes with insulation tape. Store at 4°C for the required time (usually between 4 and 8 weeks). In our experience, the inclusion of fresh silica gel is important, since if damp emulsion-coated slides are left for long periods, moulds tend to grow over the sections. Remarkably, the moulds can digest away most of the section.
8. On the day of development, slide boxes are allowed to warm up to room temperature. This prevents moisture condensing on them and possibly interfering with the emulsion. Under safe-light conditions (Kodak 6B), slides are transferred into glass racks.
9. Immerse slides in 250 ml D19 developer[c] (17°C) for 2 min.[d] After developing, immerse slides in deionized water for 30 s.
10. Transfer slides into 250 ml of a freshly prepared solution of 30% sodium thiosulfate for 2 min. You can also use Ilford Pan paper fixer (200 ml of pan fixer/litre).
11. Transfer slides into 250 ml distilled water for 2 min. Normal lights can be turned on at this point.
12. Wash in distilled water twice for 10 min.
13. Allow slides to air-dry for half an hour.
14. Sections are now ready to be stained.[e] Immerse sections in a 0.1% solution of thionin for several minutes.[f] This is a very variable and subjective step. Do not leave slides in thionin for too long – otherwise the section will be too heavily stained to see the silver grains. However, it is easy to get the stain out by placing the slides in water and 70% ethanol (see next step).
15. Transfer sections into water and then 70% ethanol to remove as much of

the stain as required. Transfer sections into 95% ethanol for several minutes. If you have taken too much stain out, move the slides back into 70% ethanol, then water and then the dye etc. and then back into water, 70% ethanol and 95% ethanol. When you are happy with the stain intensity, transfer sections into 100% ethanol for several minutes; then transfer sections into Histoclear (National Diagnostics) for several minutes.

16. Drain the sections of Histoclear and mount with glass coverslips and DPX mounting medium (BDH/Merck).

[a] Emulsion is Ilford K5 stored at 4°C. We prefer the K5 emulsion to the commonly used Kodak NTB2 emulsion, because the Ilford brand requires no aliquoting prior to use.
[b] The dipping chamber is a glass vessel about the same depth as the microscope slides. (Fig. 1.15). Amersham Pharmacia Biotech sell good dipping chambers. ("Hypercoat Dipping Vessels", catalogue number RPN39).
[c] D19 developer (Kodak) is made up from the powder exactly according to the supplier's instructions. It is stored in light-tight (dark glass) bottles at room temperature. When dissolved, D19 is a pale-straw colour. If it is excessively discoloured (dark brown), make up a fresh stock.
[d] Developer is cooled to 17°C on ice.
[e] Dr Andrew Gundlach has pointed out to us a method popularized by the Hokfelt group in Sweden. This is simply to photograph developed emulsions under water or glycerol coverslip with no counterstain. This helps to identify fibre tracts and grey matter regions when viewed under dark-field. If desired, sections can then be rephotographed following Nissl staining. This works well for regions such as the medula oblongata and nucleus of the solitary tract (A. Gundlach, personal communication).
[f] Thionin/Lauth's Violet, acetate salt (Sigma, no. T-3387) is dissolved in 0.1 M acetic acid, 0.1 M sodium acetate to make a 1% (w/v) stock solution. This stock solution is diluted ten-fold in 0.1 M acetic acid, 0.1 M sodium acetate to make a 0.1% working solution. Filter through cotton wool before use – this removes any large particles of undissolved dye, which could precipitate onto the sections. Other Nissl stains, such as toluidine blue and neutral red, are equally appropriate. While neutral red is less useful for deriving low-power images for illustration of neuroanatomy, it gives a particularly good contrast with the black silver grains, and this can be useful for both monochrome and colour photography/image analysis.

Emulsion types

Which emulsion should you use? We have tried them all. They all give artefacts from time to time. Make sure you choose an emulsion optimized for use with ^{35}S and ^{33}P (see Chapter 9, Section 9.5.3). Ilford K5 is good, but has to be melted (Protocol 1.9). Emulsions that come pre-melted are convenient. For example, hypercoat LM-1 (Amersham Pharmacia Biotech) comes ready to use, and does not require diluting; you simply have to pre-warm it. Artefacts, which can be obtained with any make of emulsion, include a high background of silver grains, and sometimes complete fogging, and various forms of crystals. It is often difficult to track down the cause. But given the length of time for which the slides will be dipped, it is best to dip some blank slides at the same time as doing the main experiment. These blank slides are put in a separate box and developed the next day; this at least ensures that there are no immediate problems. For example, if someone had switched on the light while the emulsion-coated slides were drying, you will see immediately, rather than finding out a few months afterwards. See also Chapter 9, Section 9.5.6, Controls for use with emulsion autoradiography.

1.10 Controls for ISH

Tissue sections contain surfaces and sites that trap probes and generate misleading results (Fig. 1.18). As discussed by Uhl (1987), there are no completely satisfactory controls for ISH studies. For ISH using oligonucleotides, the two best routine controls are competition hybridizations with excess concentrations of unlabelled oligonucleotides and the reproduction of identical autoradiographic patterns with several (i.e. at least two) independent oligonucleotides built to hybridize to different parts of the mRNA. These are the only two types of control we always do. Examples of these controls are illustrated in Figs 1.13D, F and H and 1.17. As seen from ISH on a horizontal rat brain section with a ^{35}S-labelled 45mer (KA-2I), the KA-2 mRNA is expressed in many brain regions, e.g. neocortex, hippocampus and granule cell layer of the cerebellum (Fig. 1.17A, Herb *et al.,* 1992). When a parallel horizontal section is hybridized with buffer containing both the same radiolabelled probe plus an excess of the unlabelled oligonucleotide, most of the signal is competed out (Fig. 1.17B). Some pattern lines remain over the meninges (the ventricle lining of the brain), and this is therefore interpreted as non-specific (Fig. 1.17B). The principle behind this competition control is that specific binding is saturable (i.e. finite), whereas non-specific binding increases linearly and, within the range of probe concentrations used in these experiments, can be regarded as "infinite". See Protocol 1.5, footnote *a*, for experimental details of setting up the competition hybridization (we routinely use a 100-fold excess of unlabelled probe). Figures 1.17C and D are autoradiographs obtained from two further distinct 40 and 45mer oligonucleotides KA-2b and KA-2c, designed to hybridize to different regions of the KA-2 mRNA. The pattern is the same as that obtained with the original KA-2 probe, confirming specificity.

Other controls are less informative for the specificity of a particular probe, but may sometimes be useful.

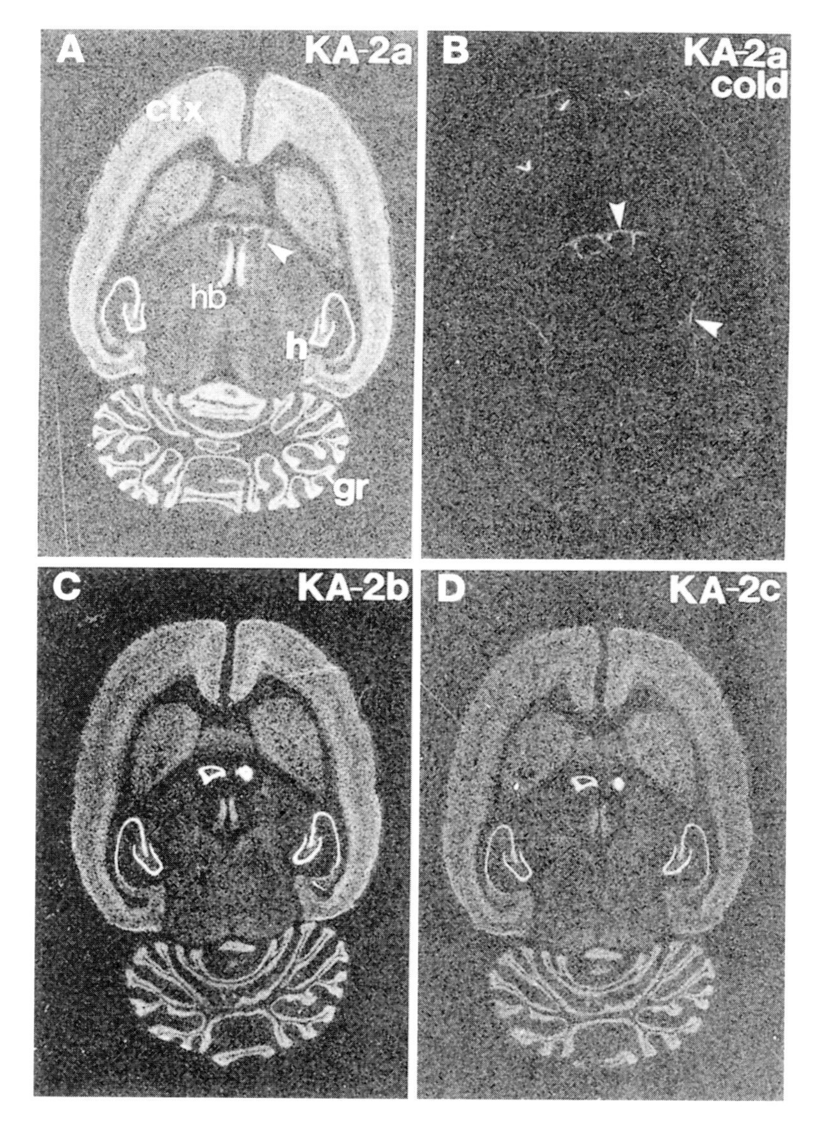

FIG. 1.17. Controls for ISH. X-ray film autoradiographic image in A resulted from hybridization of a ^{35}S-labelled KA-2a oligonucleotide to a horizontal rat brain section. The image in B was a parallel hybridization, the hybridization buffer of which contained a 100-fold excess of unlabelled KA-2a in addition to the ^{35}S-labelled KA-2a, everything else being identical. The arrowheads mark some non-specific hybridization over the ventricle linings (meninges). Images in C and D are hybridizations with oligonucleotides KA-2b and KA-2c, which hybridize to different regions of the KA-2 mRNA (see Herb *et al.,* 1992). Ctx, neocortex; h, hippocampus; hb, medial habenulae; gr, cerebellar granule cells.

1. Hybridization with a labelled sense (coding-strand) oligonucleotide with the same length, base composition and specific activity as the antisense probe. An absence of a signal with such a sense probe does not guarantee that the signal obtained with the antisense probe is specific. The antisense probe could still be cross-hybridizing to a related sequence. However, the sense probe is useful for determining general features of non-specific hybridization to the section (see Morris, 1989).
2. Another useful control, convincing although frequently not possible, is to hybridize sections from gene-targeted mice in which the exon recognized by the ISH oligonucleotide has been deleted (see fig. 1 in Hormuzdi *et al.,* 2001). This gives a general sense of background hybridization.
3. Pretreatment of the sections prior to hybridization with ribonuclease A (RNaseA) invariably destroys the signal. To do this control, sections are removed from ethanol storage and air-dried. They are then transferred into a trough containing 250 ml of 2 × SSC with 20 μg/ml RNaseA (see Sambrook and Russell, 2001, for preparation of RNaseA stocks, and Blumberg, 1987, for the safe handling of RNaseA). Sections are incubated at 37°C for half an hour. They are then dehydrated through 0.1 × SSC, 70% and 95% ethanol and hybridized with labelled probe in the standard manner (Protocol 1.5). *Important:* Glassware and solutions that come into contact with RNaseA must be kept separate from those used for the main experiments. Since the majority of non-specific hybridization is to RNA, this control is not too informative.
4. Northerns. If a probe gives a particular autoradiographic pattern after hybridizing to brain sections, it usually gives discrete band(s) on a northern blot, given that enough mRNA has been loaded and hybridization and washing conditions resemble the ISH experiment. When using an oligonucleotide on a northern membrane, the oligonucleotide is labelled by 3′-tailing with [α^{32}P]dATP (6000 Ci/mmol), then hybridized at 50% formamide/4 × SSC at 42°C and washed at 1 × SSC at 60°C (Morris, 1989; Wisden *et al.,* 1988, 1990). However, if it is not necessary to know mRNA size, ISH with multiple probes is an adequate control (assuming no alternative splicing).

For more discussion on the use of controls in ISH, see Chapter 9, Sections 9.4.7 and 9.5.6 and Chapter 11, Section 11.5.5.

PROTOCOL 1.10 OPTIONAL PRETREATMENT OF SECTIONS PRIOR TO HYBRIDIZATION[a]

For all the steps listed below, use clean glassware and sterile water to make up all solutions. Use 250 ml of each solution in BDH/Merck glass staining troughs (see Fig. 1.4)

1. Transfer sections straight from 95% ethanol into 1 × PBS.
2. Transfer into 0.25% acetic anhydride in 0.1 M triethanolomine–HCl pH 8.0/0.9% NaCl for 10 min at room temperature.[b]
3. Transfer into 70% ethanol for 1 min.
4. Transfer into 95% ethanol for 2 min.
5. Transfer into 100% ethanol for 1 min.
6. Transfer into chloroform (100%) for 5 min.
7. Transfer back through 100% (1 min) and 95% ethanol (1 min).
8. Air-dry sections and proceed to Protocol 1.5, step 1.

[a] This protocol is taken from Young *et al.* (1986a).

[b] To prepare the acetic anhydride solution, add 3.3 ml of triethanolamine (it comes as a viscous liquid), 1.25 g of NaCl and 1.0 ml of conc. HCl per 250 ml of sterile water. Mix well. Just prior to use add 625 μl of acetic anhydride. Stir well with a sterile pipette tip.

1.11 Artefacts and Failures: Troubleshooting

If you follow the conditions for ISH on brain as presented in this chapter, particularly with regard to probe concentrations, non-specific signals are usually absent. Sometimes the cerebellar granule cell layer has a non-specific component; sometimes the white matter labels non-specifically. But in spite of everything, various other artefacts can be produced (Fig. 1.18). Fortunately, some of the characteristic artefactual staining patterns shown in Fig. 1.18 can be avoided by adapting the methodology. Spots of signal distributed over tissue and slide when using ^{35}S-labelled probes (Fig. 1.18A,B) could be because the DTT has deteriorated. Fresh DTT, or an increased DTT concentration in the hybridization buffer (Protocol 1.5, note *a*) can prevent this problem. Addition of polyadenylic acid and pyrophosphate to a "minimalist" hybridization buffer (Protocol 1.8, note *a*) can reduce any non-specific binding over the tissue (Fig. 1.18A–D), particularly over cell-dense areas. The presence of edge artefacts reflects the quality of the sections being used. Extra care over the maintenance of humidity in the hybridization chamber will reduce the chances of probe drying onto the slide (Fig. 1.18 E); storage of the sections under ethanol, as opposed to within a desiccated sealed box in a freezer, generally prevents adsorption onto white matter tracts (Fig. 1.18 F,G).

The most frequent cause of probes binding non-specifically to the section

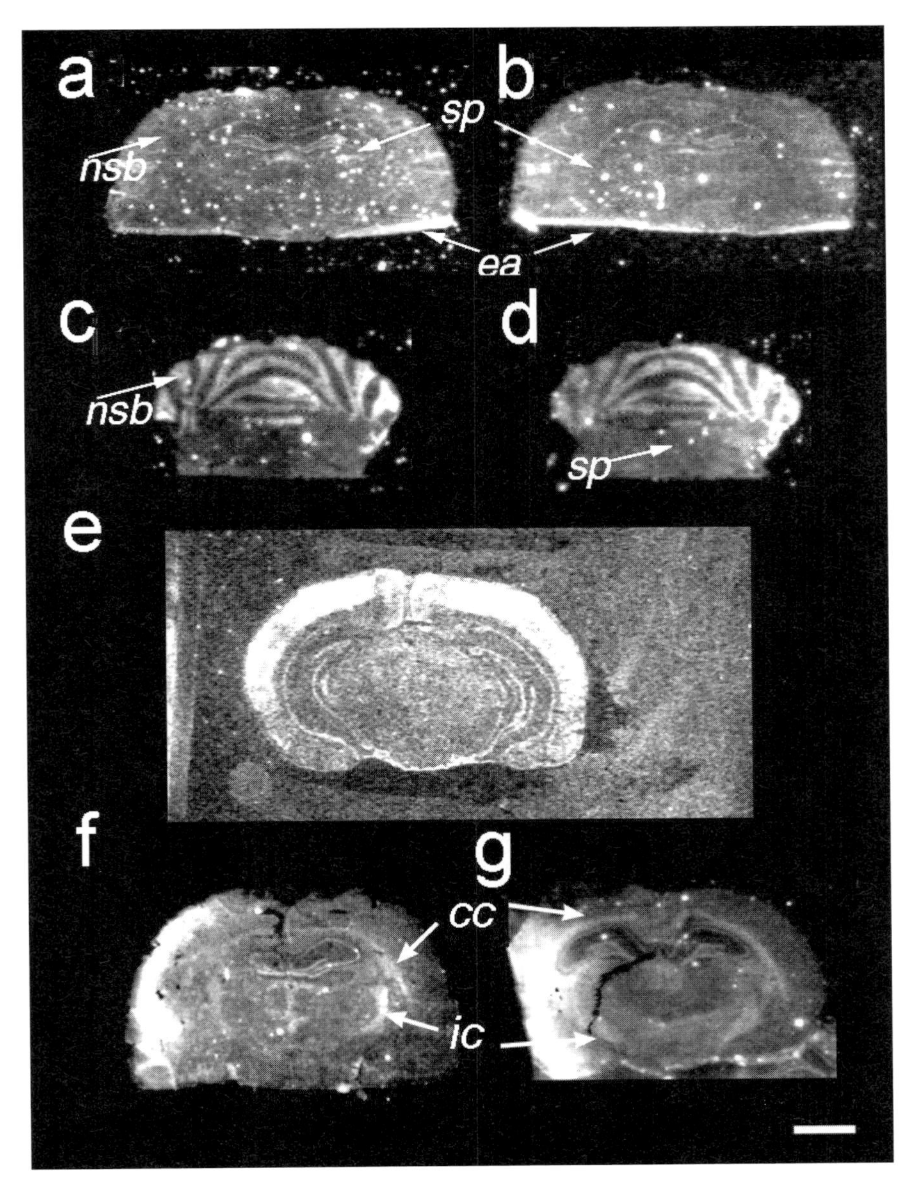

FIG. 1.18. ISH artefacts. Occasionally, characteristic artefacts appear instead of, or in addition to, specific hybridization. Spots of signal (sp) over both the tissue and the surrounding slide can be observed with ^{35}S-labelled oligonucleotide probes (A), and, as they reflect adherence to the slide by the [^{35}S]dA-tail, they are not displaced by excess unlabelled oligonucleotide (B). Such spotting can be prevented by increasing the amount of DTT in the hybridization buffer. General non-specific binding (nsb) of the labelled probe is greater in cell-dense areas such as the hippocampus and cerebellar granule cell layer (A,C). It is readily detected as it is not displaced by excess unlabelled oligonucleotide (B,D). Edge artefacts (ea) are not

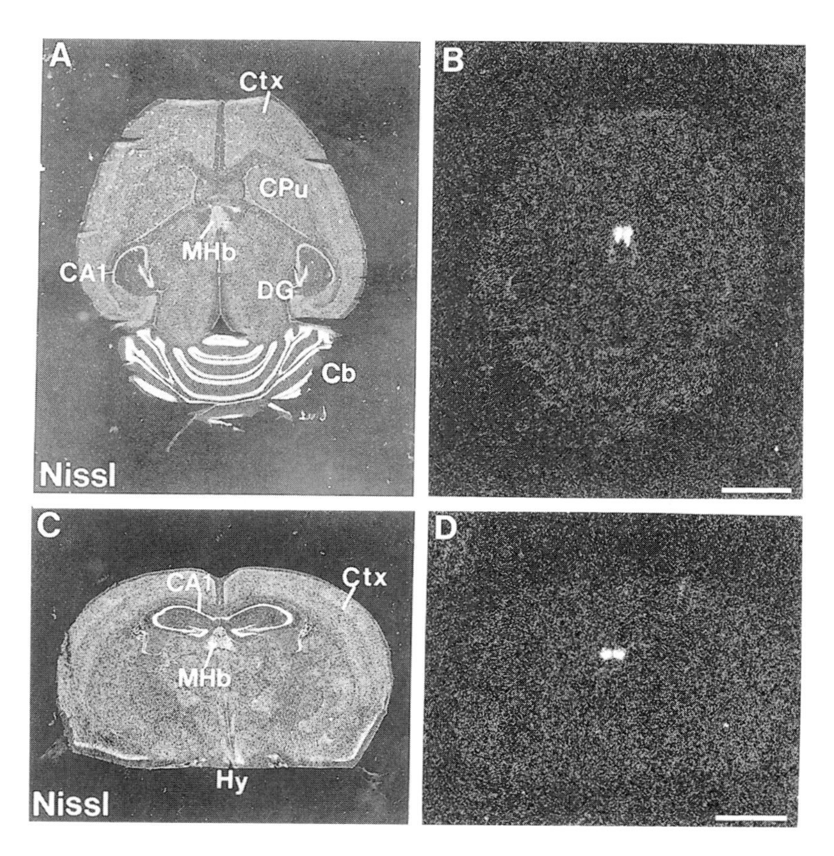

FIG. 1.19. Printing Nissl stains. X-ray film autoradiographs (B and D) were produced using an oligonucleotide hybridizing to the mRNA encoding the $5HT_{5B}$ receptor. This mRNA is mainly in the medial habenulae in adult mice (Wisden and Voigt, unpublished; Wisden *et al.*, 1993). After X-ray film exposure, the sections were thionin stained (Protocol 1.9) and the microscope slides were printed directly in the enlarger onto photographic paper (A and C), at the same magnification as used for the X-ray film, to generate an image of the entire section. Areas of high cell density appear white. These areas, such as the granule layer of the cerebellum and dentate gyrus of the hippocampus, are devoid of signal. Cb, cerebellum; CPu, caudate putamen; Ctx, neocortex; DG, dentate granule cells; Hy, hypothalamus; MHb, medial habenulae.

FIG. 1.18 *continued*

uncommon, and reflect trapping of the probe in a folded piece of tissue. If the hybridization chamber is not sufficiently humidified, or if the hybridization buffer dries onto the slide at any stage prior to washing, then there is a patchy but widespread signal over the whole slide (E). Sometimes, labelled probes adhere to white matter tracts such as the corpus callosum (cc) and internal capsule (ic), as shown (F). In contrast to specific hybridization to mRNAs in these tracts (see Fig. 1.13G), such artefactual signals are not displaced by excess unlabelled oligonucleotide (G). Scale bar 2 mm.

(e.g. Fig. 1.18A–D) is the poly([^{35}S]dA) tail being too long (e.g. counts greater than 400 000 dpm/µl of labelled probe). In this case the binding characteristics of the tail (>50 nucleotides, say) dominate the affinity of the oligonucleotide (45 nucleotides, say) for hybridization. The labelling reaction should be repeated, aiming for a specific activity within the recommended range. Artefacts from hybridizing embryo sections are shown in Fig. 3.3 (Chapter 3).

1.12 Presentation of Autoradiographs

After exposure and development, the sheets of X-ray film are cut into strips and stored in the type of plastic wallets used for protecting developed 35 mm films against scratching. Image production is straightforward. For black and white images, wet photography gives the highest quality results – sharp images with maximal contrast; however, digital photography and image production are more convenient (Section 1.12.2).

1.12.1 Wet Photography

Strips of X-ray film can be inserted directly into a photographic enlarger (just as for normal negatives), and a reverse image printed onto photographic paper (e.g. Figs 1.2, 1.8, 1.12 and 1.14). This gives a pseudo "dark-field" image, with black areas of signal on X-ray film appearing white on the printed paper. For wet photography, paper grade is critical. It is generally best to use high-contrast paper, e.g. Agfa grade 6 Rapidoprint line paper (TP6 WPt2).

It is often desirable to match an X-ray film image with its corresponding Nissl-stained section (Fig. 1.19); a convenient and rapid alternative to taking a picture of the whole section on a low-powered microscope is to insert the stained slide directly into the enlarger (or digital scanner). First the section is stained with thionin (or equivalent Nissl stain) using the protocol given in Protocol 1.9, steps 13–16, and mounted under a coverslip. After the slide has dried, dust particles are cleaned off by wiping with 70% ethanol. The X-ray film image is printed first. Then the slide is inserted into the enlarger and printed at exactly the same magnification as that used for the X-ray film. The resulting image will have regions of white corresponding to the blue Nissl stain (Fig. 1.19). This technique gives good results for vertebrate embryo sections (Fig. 3.1, Chapter 3), and also works for printing sections

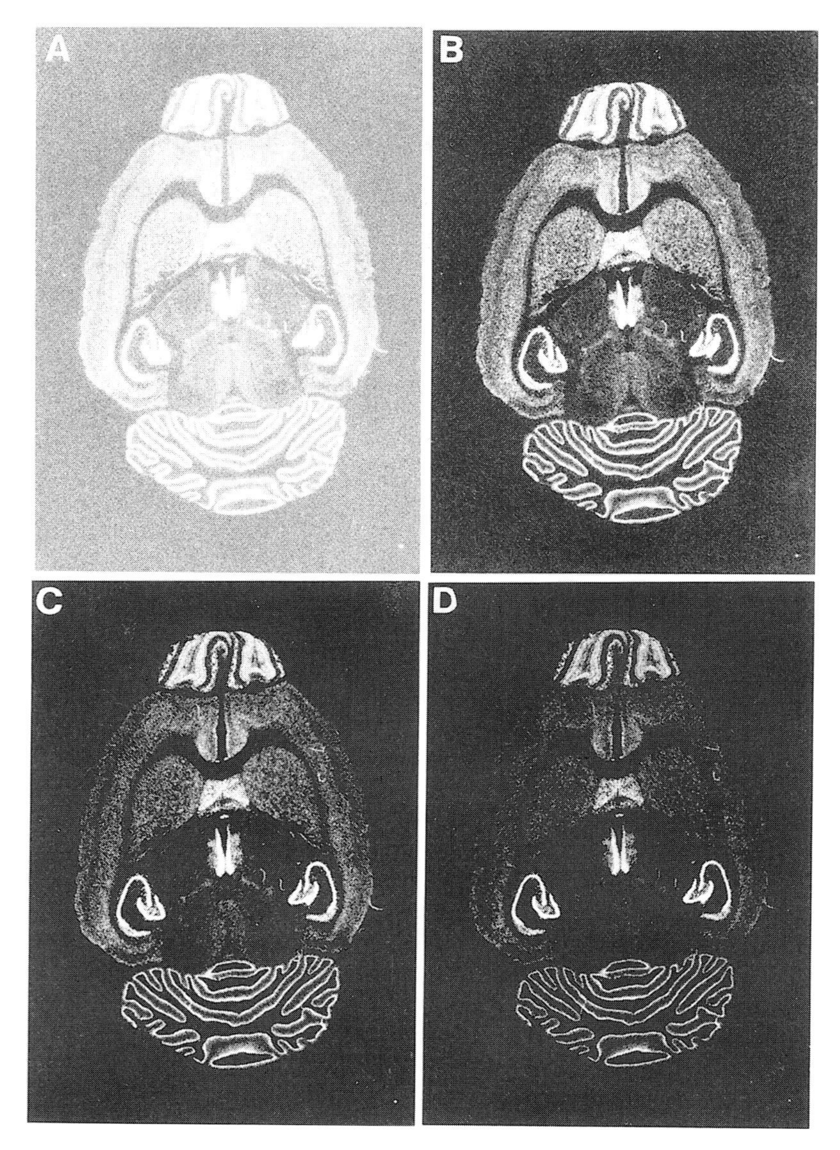

FIG. 1.20. Presenting X-ray film autoradiographs. The same image of the AMPA-kainate receptor GluR-A subunit mRNA distribution in rat brain (Keinänen *et al.*, 1990) is in all four photographs, but exposure time of light to the photographic paper differed by 5-s increments. The same principle applies to images manipulated electronically. Oligonucleotide for GluR-A subunit mRNA was as described in Kainänen *et al.* (1990).

hybridized with non-radioactive probes (see Figs 10.7 and 10.8 and Chapter 11, Section 11.7 and Fig. 11.3A,B).

1.12.2 Digital Photography

Digital methods produce results that are almost as good as those from wet photography, in far less time (Fig. 1.3). Standard digital protocols involve two stages:

1. Capturing the autoradiographic image, either from an autoradiographic film displayed on a lightbox, or from an emulsion-coated slide digitally photographed under the microscope. The image capture quality is the major factor determining the quality of the final image, so the highest resolution camera available should be used. Two useful books on digital photography are by Busch (2000) and Greenberg (1999).
2. Importing the digital image into an image-processing software package, such as Adobe Photoshop (www.Adobe.com). If the figure is to be a montage of different images, then these are also imported and added, and the figure is then labelled (see Figs 1.3 and 1.13). The image can then be manipulated in many ways – trimming, optimizing brightness and contrast; and displaying as bright-field (Fig. 1.3) or dark-field (Figs 1.13 and 1.18). Increasing the contrast can strongly improve the appearance of an autoradiographic image. The final electronic figure can be printed on photographic-quality paper, or exported into presentation software, such as Microsoft Powerpoint. Two good books on Adobe Photoshop are from Shufflebotham (2001) and Stanley (2001).

1.12.3 Subjectivity

As is the case for choosing the correct exposure time of the primary autoradiograph (how long to leave a slide exposed to X-ray film or dipped in emulsion), turning autoradiographs into images with the correct contrast is subjective. It depends on how much or how little signal you want to see (Chapter 9, Section 9.4.8). The mRNA distribution can be made to change vividly depending on printing or electronic scanning conditions. Figure 1.20 illustrates this. The same X-ray film autoradiograph (AMPA receptor GluR-A subunit mRNA in the rat brain) produced all four pictures, but exposure time in the enlarger varied by 5-s increments (the same would apply to electronic versions). Image A (Fig. 1.20) is definitely underexposed ("underprinted") with the mRNA seemingly abundant everywhere. But without knowledge of the primary autoradiograph, it is unclear whether B, C or D is the "correct" distribution. It is, in fact, image B. Image D is extreme and is "overprinted" with the GluR-A gene apparently expressed mainly in the hippocampus and habenulae, with little neocortical expression. Clearly, there can be a fine line between justifiable enhancement of an image and distortion. Particular care is needed when a composite figure illustrates differences in hybridization signal between images that have been processed separately. It is then essential to manipulate each image within the figure identically (optimization of brightness and contrast).

Acknowledgements

We are very grateful to Ulla Amtmann, who read the manuscript, and did some of the experiments illustrated. WW thanks Professors Stephen P. Hunt and Peter H. Seeburg for their initial help and encouragement. BJM is supported by the Biotechnology and Biological Sciences Research Council (UK); WW is supported by the Deutsche Forschungs Gemeinschaft (Germany). Dr C. McKerchar (YRING, University of Glasgow) provided the autoradiograms for Fig. 1.13 (B,G,H); Dr Isabel Aller provided Fig. 1.6; Dr Sean Munro (MRC Laboratory of Molecular Biology, Cambridge, England) suggested using glass beads in the spin columns. We thank the people that contacted us over the last 7 years with queries about the protocols. Their comments helped us strengthen the text.

References

Albretsen, C., Haukanes, B. I., Aasland, R., and Kleppe, K. (1988). *Anal. Biochem.* **170,** 193–202.

Barton, A. J., Pearson, R. C., Najlerahim, A., and Harrison, P. J. (1993). *J. Neurochem.* **61,** 1–11.

Berthele, A., Laurie D. J., Platzer, S., Zieglegansberger, W., Tölle, T. R., and Sommer, B. (1998). *Neuroscience* **85,** 733–749.

Berthele, A., Boxall S. J., Urban, A., Anneser, J. M., Zieglgänsberger, W., Urban, L., and Tölle. T. R. (1999). *Dev. Brain Res.* **112,** 39–53.

Blumberg, D. D. (1987). *Methods Enzymol.* **152,** 20–24.

Branks, P. L. and Wilson, M. C. (1986). *Brain Res.* **387,** 1–16.

Busch, D. D. (2000). "Digital Photography for Dummies: Quick Reference". IDG Books Worldwide, Foster City, CA.
Campos, M. L., de Cabo, C., Wisden, W., Juiz, J. M., and Merlo, D. (2001). *Neuroscience* **102,** 625–638.
Davidson, D. and Baldock, R. (2001). *Nature Rev. Genet.* **2,** 409–417.
Davidison, D., Bard, J., Kaufman, M., and Baldock, R. (2001). *Trends Genet.* **17,** 49–51.
Emson, P. C. (1993). *Trends Neurosci.* **16,** 9–16.
Erdtmann-Vourliotis, M., Mayer, P., Riechert, U., Händel, M., Kriebitzsch, J., and Höllt, V. (1999). *Brain Res. Protocols* **4,** 82–91.
Eschenfeldt, W. H., Puskas, R. S., and Berger, S. L. (1987). *Methods Enzymol.* **152,** 337–342.
Garner, G. C., Tucker, R. P., and Matus, A. (1988). *Nature* **336,** 674–677.
Gilmore, J. H., Lawler, C. P., Eaton, A. M., and Mailman, R. B. (1993). *Mol. Brain Res.* **18,** 290–296.
Glencorse, T. A., Bateson, A. N., and Darlison, M. G. (1992). *Eur. J. Neurosci.* **4,** 271–277.
Greenberg, S. (1999). "The Complete Idiot's Guide to Digital Photography". Alpha Books Que, Indianapolis.
Hall, J., Thomas, K. L., and Everitt, B. J. (2001). *J. Neurosci.* **21,** 2186–2193.
Harrison, P. J., Heath, P.R., Eastwood, S. L., Burnet, P. W., McDonald, B., and Pearson R. C. (1995). *Neurosci. Lett.* **200,** 151–154.
Herb, A., Burnashev, N., Werner, P., Sakmann, B., Wisden, W., and Seeburg, P. H. (1992). *Neuron* **8,** 775–785.
Herb, A., Wisden, W., Catania, M. V., Marachel, D., Dresse, A., and Seeburg, P. H. (1997). *Mol. Cell Neurosci.* **8,** 367–374.
Higuchi, M., Maas, S., Single, F. N., Hartner, J., Rozov, A., Burnashev, N., Feldmeyer, D., Sprengel, R., and Seeburg, P. H. (2000). *Nature* **406,** 78–81.
Hormuzdi, S. G., Pais, I., LeBeau, F. E., Towers, S. K., Rozov, A., Buhl, E. H., Whittington, M. A., and Monyer, H. (2001). *Neuron* **31,** 487–495.
Johnston, H. M. and Morris, B. J. (1994). *J. Neurochem.* **63,** 379–382.
Johnston, H. M. and Morris, B. J. (1995). *Brain Res. Mol. Brain Res.* **31,** 141–150.
Jones, A., Bahn, S., Grant, A. L., Köhler, M., and Wisden, W. (1996). *J. Neurochem.* **67,** 907–916.
Keinänen, K., Wisden, W., Sommer, B., Werner, P., Herb, A., Verdoorn, T. A., Sakmann, B., and Seeburg, P. H. (1990). *Science* **249,** 556–560.
Kingsbury, A. E., Foster, O. J. F., Nisbet, A. P., Cairns, N., Bray, L., Eve, D. J., Lees, A. J., and Marsden, C. D. (1995). *Mol. Brain Res.* **28,** 311–318.
Lathe, R. (1985). *J. Mol. Biol.* **183,** 1–12.
Laurie, D. J. and Seeburg, P. H. (1994). *J. Neurosci.* **14,** 3180–3194.
Laurie, D. J., Seeburg, P. H., and Wisden, W. (1992a). *J. Neurosci.* **12,** 1063–1076.
Laurie, D. J., Wisden, W., and Seeburg, P. H. (1992b) *J. Neurosci.* **12,** 4151–4172.
Laurie, D. J., Putzke, J., Zieglgänsbeger, W., Seeburg, P.H., and Tölle, T.R. (1995). *Mol. Brain Res.* **32,** 94–108.
Lewis, M. E., Sherman, T. G., and Watson, S. J. (1985). *Peptides* **6** (Suppl. 2), 75.
Lewis, M. E., Krause, R. G., and Roberts-Lewis, J. M. (1988). *Synapse* **2,** 308.
Malosio, M. L., Marqueze-Pouey, B., Kuhse, J., and Betz, H. (1991). *EMBO J.* **10,** 2401–2409.
Marvanová, M., Törönen, P., Storvik, M., Lasko, M., Castrén, E. and Garry Wong (2002). Mol. Brain Res. (In Press)
Miralles, C. P., Gutierrez, A., Khan, Z. U., Vitorica, J., and De Blas, A. L. (1994). *Brain Res. Mol. Brain Res.* **24,** 129–139.
Monyer, H., Seeburg, P. H., and Wisden, W. (1991). *Neuron* **6,** 799–810.
Morris, B. J. (1989). *J. Comp. Neurol.* **290,** 358–368.

Morris, B. J. (1995). *J. Biol. Chem.* **270,** 24740–24744.
Morris, B. J. (1997). *Eur. J. Neurosci.* **9,** 2334–2339.
Morris, B. J., Feasey, K. J. ten Bruggencate, G., Herz, A., and Höllt, V. (1988). *Proc. Natl Acad. Sci.* (USA) **85,** 3226–3230.
Morris, B. J., Haarmann, I., Kempter, B., Höllt, V., and Herz, A. (1986). *Neurosci. Lett.* **69,** 104–108.
Morris, B. J., Hicks, A. A., Wisden, W., Darlison, M. G., Hunt, S. P., and Barnard, E. A. (1990). *Mol. Brain Res.* **7,** 305–315.
Mulhardt, C., Fischer, M., Gass, P., Simon-Chazottes, D., Guenet, J. L., Kuhse, J., Betz, H., and Becker, C. M. (1994). *Neuron* **13,** 1003–1015.
Paterlini, M., Revilla, V., Grant, A. L., and Wisden, W. (2000). *Neuroscience* **99,** 205–216.
Persohn, E., Melherbe, P., and Richards, J.G. (1992). *J. Comp. Neurol.* **326,** 193–216.
Pittius, C.W., Kley, N., Loeffler, J.P., and Höllt, V. (1985). *EMBO J.* **4,** 1257–1260.
Pratt, G. D. and Kokaia, M. (1994). *Trends Pharmacol. Sci.* **15,** 131–135.
RIKEN Genome Exploration Research Group Phase II Team and the FANTOM Consortium (2001). *Nature* **409,** 685–690.
Roberts, L. A., Large, C. H., Higgins, M. J., Stone, T. W., O'Shaughnessy, C. T., and Morris, B. J. (1998). *Mol. Brain Res.* **56,** 38–44.
Rogers, A. W. (1979). "Techniques of Autoradiography". Elsevier, New York
Ross, B. M., Knowler, J.T., and McCulloch, J. (1992). *J. Neurochem.* **58,** 1810–1819.
Sambrook, J. and Russell, D. W. (2001). "Molecular Cloning: A Laboratory Manual", third edition. Cold Spring Harbor Laboratory Press, Cold Spring Harbor, New York.
Shufflebotham, R. (2001). "Photoshop 6 in Easy Steps". Computer Step, Southam.
Simpson, C. S. and Morris, B. J. (2000). *J. Biol. Chem.* **275,** 16879–16884.
Sinkkonen, S. T., Hanna, M. C., Kirkness, E. F., and Korpi, E. R. (2000). *J. Neurosci.* **20,** 3588–3595.
Sommer, B., Kainänen, K., Verdoorn, T. A., Wisden, W., Burnashev, N., Herb, A., Köhler, M., Takagi, T., Sakmann, B., and Seeburg, P. H. (1990). *Science* **249,** 1580.
Stanley, R. (2001). "The Complete Idiot's Guide to Adobe Photoshop 6". Alpha Books, Que, Indianapolis.
Sugino, H., Hamada, S., Yasuda, R., Tuji, A., Matsuda, Y., Fujita, M., and Yagi, T. (2000). *Genomics* **63,** 75–87.
Thomas, K. L. and Everitt, B. J. (2001). *J. Neurosci.* **21,** 2526–2535.
Thomas, K. L., Laroche, S., Errington, M. L., Bliss, T. V., and Hunt, S. P. (1994). *Neuron* **13,** 737–745.
Thomas, K. L., Davis, S., Hunt, S. P., and Laroche, S. (1996). *Learn Mem* **3,** 197–208.
Tölle, T. R., Berthele, A., Zieglgänsberger, W., Seeburg, P. H., and Wisden, W. (1993). *J. Neurosci.* **13,** 5009–5028.
Trivedi, P. G., Yu, H., Trumbauer, M., Chen, H., Van der Ploeg, L. H., and Guan, X. (2001). *Peptides* **22,** 395–403.
Uhl, G. R. (Ed.) (1987). "*In Situ* Hybridization in Brain". Plenum Press, New York.
Uhl, G. R., Cwickel, B., Pagnonis, C., and Habener, J. (1985). *Annal. Neurol.* **18,** 149.
Wahl, G. M., Berger, S. L., and Kimmel, A. R. (1987). *Methods Enzymol.* **152,** 399–407.
Wisden, W. and Seeburg, P. H. (1993). *J. Neurosci.* **13,** 3582–3598.
Wisden, W., Morris, B. J., Darlison, M. G., Hunt, S. P., and Barnard, E. A. (1988). *Neuron* **1,** 937–947.
Wisden, W., Errington, M. L., Williams, S., Dunnett, S. B., Waters, C., Hitchcock, D., Evan, G., Bliss, T. V. P., and Hunt, S. P. (1990). *Neuron* **4,** 603.
Wisden, W., Morris, B. J., and Hunt, S. P. (1991). *In* "Molecular Neurobiology: A Practical Approach" (J. Chad and H. Wheal, Eds), pp.205–225. IRL Press/Oxford University Press, Oxford.
Wisden, W., Laurie, D. J., Monyer, H., and Seeburg, P. H. (1992). *J. Neurosci.* **12,** 1040–1062.

Wisden, W., Parker, E. M., Mahle, C. D., Grise, D. A., Nowak, H. P., Yocca, F. D., Felder, C. C., Seeburg, P. H.. and Voigt, M. M. (1993). *FEBS Lett.* **333:** 25–31.
Wisden, W., Seeburg, P. H., and Monyer, H. (2000). *In* "Handbook of Chemical Neuroanatomy, Vol. 18 Glutamate" (O. P. Ottersen and J. Storm-Mathisen, Eds), pp.99–143. Elsevier, Amsterdam.
Wu, Q. and Maniatis, T. (1999). *Cell* **97,** 779–790.
Young, W. S. III (1989). *Methods Enzymol.* **168,** 702–709.
Young, W. S. III, Mezey, E., and Siegel, R. E. (1986a). *Neurosci. Lett.* **70,** 198–203
Young, W. S. III, Bonner, T. I., and Brann, M. R. (1986b). *Proc. Natl Acad. Sci. USA* **83**, 9827.
Zirlinger, M., Kreiman, G., and Anderson, D. J. (2001). *Proc. Natl Acad. Sci. USA* **98,** 5270–5275.

CHAPTER 2

CRYOSTAT SECTIONING OF BRAINS

V. Revilla[1] and A. Jones

MRC Laboratory of Molecular Biology, MRC Centre, Hills Road, Cambridge CB2 2QH, UK

[1]Present address: Almirall-Prodespharma, Cardener 68-74, 08024 Barcelona, Spain

2.1 Introduction

Cutting good sections on a cryostat can be the most frustrating part of any *in situ* hybridization protocol. Here we provide a description of the essential features of a cryostat, and general advice on troubleshooting (Section 2.6). Details and peculiarities of cutting specific tissues are given in other chapters. Cutting matching sections for quantification is described in Chapter 9, Table 9.1, notes *a–d*. Cutting paraffin-embedded sections on a cryostat for combined *in situ* hybridization and immunocytochemistry is discussed in Chapter 11, Section 11.3.2 and cutting fixed tissue in Section 11.4.

2.2 Parts of a Cryostat

There are two main manufacturers of cryostats: Leica Microsystems Nussloch GmbH (www.leica.com/specimen-prep) and Bright Instruments Co. Ltd (www.brightinstruments.com). Both make open-top cryostats in a variety of models ranging from the simplest non-motorized sectioning to the computerized cryomicrotome from Leica.

INTERNATIONAL REVIEW OF NEUROBIOLOGY, VOL. 47
ISSN/02 $00.00

The core of a cryostat sectioning mechanism is depicted in Fig. 2.1. The main parts are as follows:

A

B

FIG. 2.1. Parts of a cryostat. (A) Bright model. (B) Leica model.

1. Knife holder and anti-roll guide plate. Knife holders can accommodate both reusable knives (Figs 2.1A and 2.4) and/or disposable blades (Figs 2.1B and 2.5A). The anti-roll guide plate (Fig. 2.6) is usually made of glass or plastic and it is probably the most important adjunct to the preparation of good sections.
2. Specimen holder and orientation. The specimen disc (with the specimen mounted onto it) fits directly into the object holder and it usually has screws for tightening (Fig. 2.1A,B) and changing orientation. Some models of cryostat have additional screws for secure tightening and two-dimensional orientation of the specimen holder (Fig. 2.3B).
3. Section thickness and hand wheel. The thickness of the sections can be adjusted using a knob (Fig. 2.1A,B). The wheel allows the movement of the specimen holder downwards so it is possible to obtain sections (the tissue section should pass smoothly between the knife and the anti-roll guide plate). The operation of the hand wheel can be manual or automated, depending on the cryostat model.

2.3 Tissues

For cryostat sectioning it is important to avoid water crystals forming in the tissue when the tissue is frozen. For this you can use liquid nitrogen or dry ice, although we prefer dry ice since with liquid nitrogen the brain cracks easily. It is important to keep the original shape of the brain when freezing to allow easy identification of the anatomical structures. For this we use a Petri dish covered with aluminium foil or a flat metallic surface previously cooled to the dry ice temperature (Fig. 2.2). Once completely frozen, the brain can be sectioned directly, after equilibrating to the cryostat chamber temperature, or stored at –80°C. When kept at –80°C, it is important to prevent dehydration of the brain by covering it with parafilm to form an airtight parcel and then putting it into a tightly closed container. A desiccated brain will give sections that crumble. Note: straight after dissection and before freezing, brains can be pre-cut with a razor blade down the sagittal, horizontal or coronal plane. This makes them easier to cut on the cryostat.

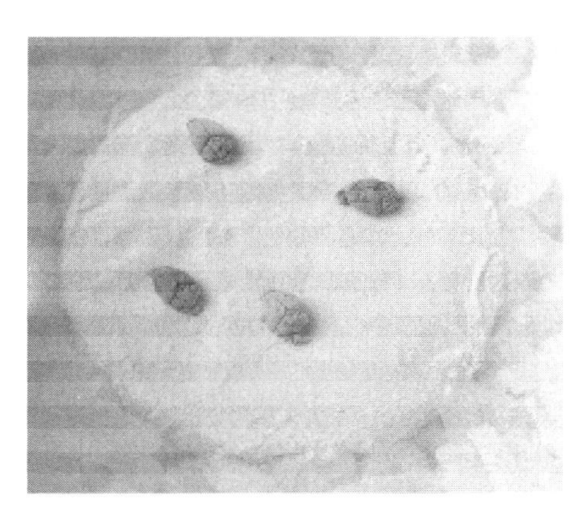

FIG. 2.2. Brains frozen on dry ice, supported on an aluminium foil-covered Petri dish.

2.4 Mounting Brain on Specimen Disc

Specimen discs come in a variety of sizes and forms (Fig. 2.3A,B) depending on the model of cryostat. The usual mounting medium (Cryo-M-Bed; Bright Instrument Co.) allows the correct adhesion of the brain to the disc, which is very important for obtaining sections of consistent thickness. Squeeze a few drops of mounting medium onto a room temperature disc, chill the disc on dry ice or in the cryostat chamber and when the medium is nearly frozen insert the brain (Fig. 2.3A). The brain should then be inserted into the mounting medium in different ways depending on the desired orientation of the sections: coronal (Fig. 2.3B), horizontal (Fig. 2.3C) or sagittal (Fig. 2.3D). Mounted tissue should be left inside the cryostat chamber for 10–15 min to allow the specimen to equilibrate to chamber temperature, otherwise the sections will crack.

2.5 Sectioning

The following hints apply to all makes of cryostat.

1. First check that the chamber, blade, anti-roll guide, specimen disc and brain are at the right temperature for sectioning. For adult brain this is normally between –18 and –20°C; when sectioning embryos or brains from very young animals (see Chapter 3), it is

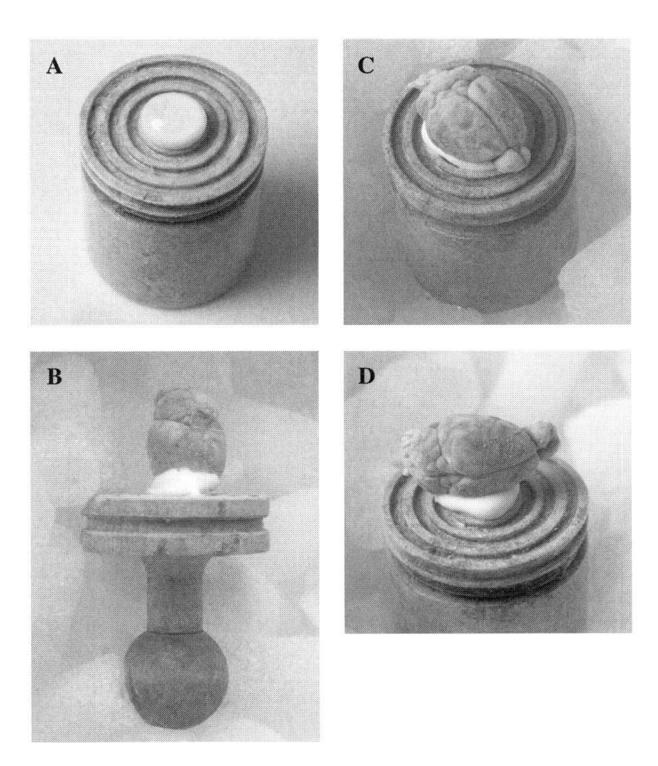

FIG. 2.3. Mounting brains. (A) Mounting medium on chilled specimen disc; brain mounted for coronal (B), horizontal (C) and sagittal (D) sectioning. See Chapter 5, Fig. 5.1c for spinal cord mounting.

convenient to raise the temperature to –12 or –13°C. Then choose the desired thickness: 14–15 μm for radioactive *in situ* hybridization as described in Chapter 1.

2. Check that the blade is in good condition. Reusable knives can be moved to expose a sharp, undamaged cutting edge but will need to be re-sharpened every now and then. This is not a problem with disposable blades, but take care since some batches of these blades come in poor-quality stainless steel, which can make sectioning very troublesome. In a busy lab, where many people are cutting, it is a good idea if people have their own blades; blades are stored at –20°C. Make sure that the blade is clean; for this use a thick artists' brush and brush the blade (Fig. 2.5G) in an upward motion to avoid damaging the blade edge.

FIG. 2.4. Cryostat set up for cutting sections. Notice the position of brain, knife and anti-roll guide plate.

3. Check that the anti-roll guide plate is correctly positioned with respect to the blade. If it is incorrectly positioned, the problems shown in Fig. 2.6 can occur. For adjustment, cryostats have a screw, normally at the base of the anti-roll guide plate (Fig. 2.1A,B). It is also important that the edge of the anti-roll guide is not damaged (if so it is normally possible to buy a new guide from your cryostat supplier) and that it is clean. Brain is a very greasy tissue so it is important to clean the guide with an alcoholic solution, otherwise sections will stick to the guide. Normally, cryostat manufactures sell this solution; otherwise 70% ethanol will do (provided that the guide is not made from metacrylate!). After cleaning the anti-roll guide you will have to chill it again, for this you can use dry ice.
4. Check the "clearance angle" (Fig. 2.1A). This is the angle at which the blade holder meets the specimen when sectioning. The general rule is: the harder the tissue, the larger the clearance angle and the stronger the compression inflicted in the specimen (and therefore it is more likely that cracked sections will be obtained). Brain is a soft tissue so normally a clearance angle of 0° is good enough.
5. Check that the specimen-holder block is in its rearmost position. The sectioning mechanism involves the specimen moving towards the blade at a distance given by the thickness of the sections. If the specimen holder is in its most forward position, it will not advance and you will get no section.

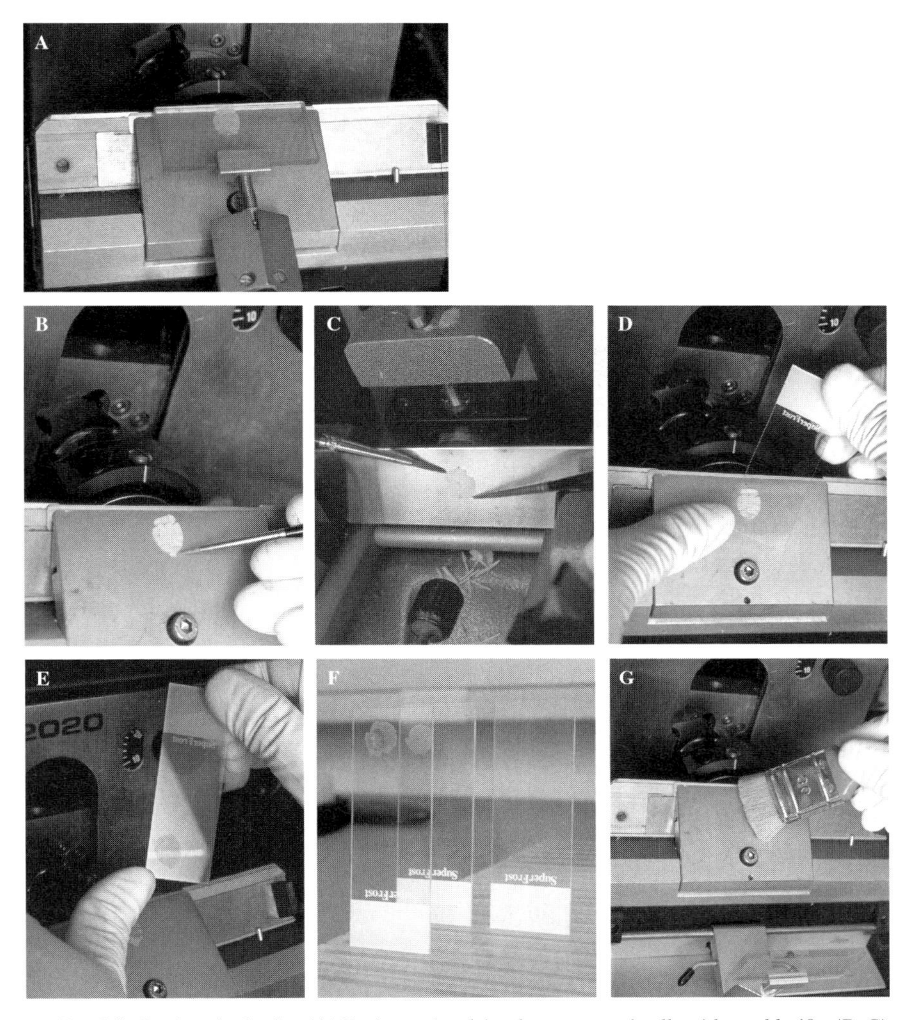

FIG. 2.5. Cutting the brain. (A) Brain section lying between anti-roll guide and knife. (B, C) Uncurling the section with fine artists' brushes before picking it up. (D, E) Picking section onto slide. (F) Sections drying on slides. (G) Cleaning the knife with a thick brush before cutting another section.

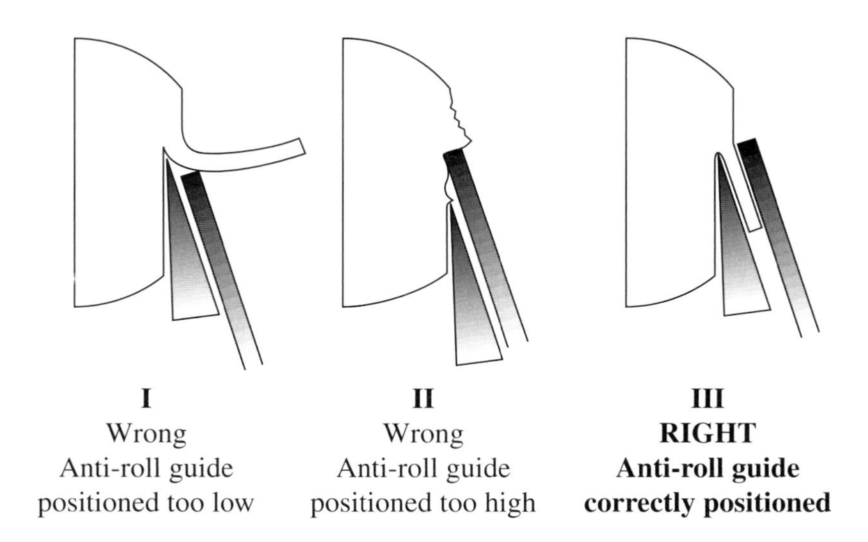

Fig. 2.6. Adjusting the anti-roll guide.

Once you have checked all this, you can start making your sections:

1. Position the specimen holder up using the handwheel (manually or motorized) and the anti-roll guide against the blade (Fig. 2.4).
2. Operate the cutting handwheel (generally a slow stroke gives better results). The tissue section should pass smoothly between the knife and the anti-roll guide plate (Fig. 2.5A).
3. Lift the anti-roll guide plate and pick up the tissue using a slide at room temperature. There should be no need to press the slide onto the section, instead the section should move onto the slide (Fig. 2.5D).
4. Often the section is curled up or creased. You can get rid of the creases using one or two fine brushes. Sometimes when it is very difficult to uncurl the sections it helps to hold the section against the blade for a few seconds using two brushes (Fig. 2.5B, C).
5. Before making further sections clean the blade and the anti-roll guide with the thick artists' brush (Fig. 2.5G).
6. Allow the sections to dry at room temperature before proceeding to fix them for the *in situ* hybridization (Fig. 2.5F). (Protocol 1.2, Chapter 1).

2.5.1 Notes

While the sections are still moist on the slide, the detailed anatomical level of the sections can be checked using a low-power stereomicroscope with a dark-field or transillumination base; white matter and grey matter tracts can be clearly seen and matched to an atlas.

It is usually best to place multiple sections on the same slide (see, for example, the left-hand column in Fig. 1.11, Chapter 1 and Fig. 5.1, Chapter 5). This is especially important for quantitative experiments (see Chapter 5, Section 5.3.2 and Chapter 9, Table 9.1, footnotes *a–d*).

2.6 Troubleshooting Guide

Cutting on the cryostat can be frustrating; some days are more difficult than others. Below are listed some common problems and their solutions.

1. Variable section thickness
   - Mounting medium does not adhere firmly to dirty specimen disc: clean disc with manufacturer's cleaning solution or 70% ethanol, re-chill holder on dry ice and re-mount specimen.
   - Specimen disc or knife not firmly clamped: tighten appropriate screws.
   - Ice on the knife advancing mechanism: defrost and clean cryostat.
   - External vibrations, e.g. from building work: wait until vibrations stop.
   - Knife blunt: replace disposable blade or move knife to expose sharp cutting area.
   - Specimen warmer or colder than cryostat chamber: leave specimen in chamber for 10–15 min for temperature equilibration.
   - Poor-quality disposable blades distort in cold.
2. Horizontal scores in sections
   - Tissue too cold: leave for further 10–15 min to equilibrate to chamber temperature.
   - Tissue dehydrated: store frozen tissue wrapped in an airtight parafilm package.
   - Tissue frozen too slowly so ice crystals form: freeze brains rapidly on dry ice (Fig. 2.3).

3. Vertical scores in sections
   - Damaged blade: replace disposable blade, move knife to an undamaged cutting area.
   - Top edge of anti-roll guide plate damaged: replace plate.
   - Tissue or ice crystals stuck to blade or anti-roll plate: clean with artist's brush (Fig. 2.5G) or for more ingrained tissue residue, clean as above.
4. Sections stick to anti-roll guide plate
   - Dirty anti-roll plate: clean as above.
   - Anti-roll guide plate too warm: cool on dry ice.
   - Badly adjusted anti-roll plate: re-adjust (Fig. 2.6).
5. Sections curl up
   - Warm air currents in the chamber: gently uncurl and flatten sections using a fine artist's brushes (Fig. 2.5B, C).
6. Sections compress
   - Anti-roll guide plate badly positioned: reposition (Fig. 2.6).
7. Specimen not advancing to knife
   - Thickness control set to zero.
   - Microtome feed screw at end of travel (Bright models): rewind.
   - Control spindle iced up: defrost cryostat.
8. Problems with ambient temperature
   - Sections difficult to pick up as glass slide too cold: warm up slides, e.g. using small electric blanket.
   - Room too hot and/or too humid so sections stick to anti-roll guide or knife: move cryostat to air-conditioned room; if not available, then stop sectioning, leave the machine to recover and meanwhile chill the anti-roll guide and knife on dry ice.

CHAPTER 3

PROCESSING RODENT EMBRYONIC AND EARLY POSTNATAL TISSUE FOR *IN SITU* HYBRIDIZATION WITH RADIOLABELLED OLIGONUCLEOTIDES

D. J. Laurie,* P. C. U., Schrotz,** H. Monyer,† U. Amtmann†

*DRA Oncology, Novartis Pharma AG, CH-4002 Basel, Switzerland
**Department of Neurobiology, University of Heidelberg, Im Neuenheimer Feld 364, 69120 Heidelberg, Germany
†Department of Clinical Neurobiology, University of Heidelberg, Im Neuenheimer Feld 364, 69120 Heidelberg, Germany.

3.1 Introduction

The expression pattern of a gene during development is a valuable piece of information (Ringwald *et al.*, 1994; Williams, 2000; Davidson and Baldock, 2001). Reliable and presentable data require good-quality sections from accurately aged animals (Fig. 3.1). Practice and care are necessary to dissect, manipulate and prepare the delicate tissues from these stages without damage or distortion. This chapter describes steps and hints useful for preparing embryonic and early postnatal (P0–P20) tissue for the *in situ* hybridization protocol with radiolabelled oligonucleotides (Chapter 1). This method has been used by us and associated colleagues to study the developmental (embryonic and postnatal) expression of many genes: $GABA_A$ receptor subunits (Laurie *et al.*, 1992); NMDA receptor subunits and RNA splice forms (Lomeli *et al.*, 1993; Laurie and Seeburg, 1994; Monyer *et al.*, 1994; Nase *et al.*, 1999);

ISSN/02 $00.00

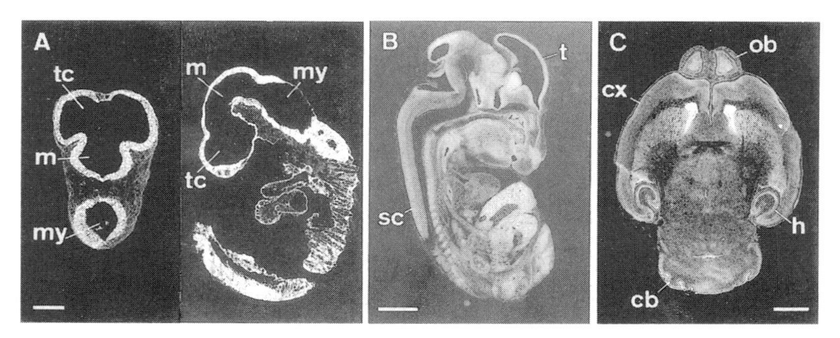

FIG. 3.1. Nissl-stained embryo sections. (A) Prefixed mouse E9.5 embryos in transverse (left) and parasagittal (right) sections, (B) E14 rat embryo (sagittal), and (C) rat P0 postnatal brain (horizontal). cb, cerebellum; cx, cortex; h, hippocampus; m, mesecoele; my, myelocoele; ob, olfactory bulb; sc, spinal cord; t, telencephalon; tc, telocoele. Scale bars 0.5 mm (A) and 2 mm (B and C).

AMPA receptor subunits and splice forms (Monyer *et al.*, 1991, 1994); kainate receptor subunits (Bahn *et al.*, 1994); serotonin receptors (Voigt *et al.*, 1991; Wisden *et al.*, 1993); peptides (Ryan *et al.*, 1997); connexin channel subunits (Hormuzdi *et al.*, 2001); peptide receptors (Burazin and Gundlach, 1999; Burazin *et al.*, 2000); netrin and its receptors (Livesey and Hunt, 1997); ion cotransporters (Giffard *et al.*, 2000); hox, and basic helix-loop-helix transcription factors (Greene *et al.*, 1998; Sinclair *et al.*, 1999).

Following on from the genome sequencing projects, we believe this particular method with radiolabelled oligonucleotides is ideally suited for large-scale mapping of gene expression during development (Davidson and Baldock, 2001; Baldock *et al.*, 2001; Davidson *et al.*, 2001). The X-ray film method permits a large number of gene expression patterns to be simultaneously compared (Laurie *et al.*, 1992; Laurie and Seeburg, 1994). Even at the level of X-ray film analysis, quite detailed information on changes in gene expression can be gleaned (Fig. 3.2). For small embryos (less than age E10) whole-mount *in situ* hybridization with digoxigenin riboprobes is more appropriate. This is covered in Chapter 12. However, whole-mount *in situ* hybridization only works reliably until mouse age E10.5; after this age, probe penetration is difficult (L. Ariza-McNaughton, personal communication).

3.2 Standardizing Animal Age

Mouse and rat have a short gestation period (20–21 days). The embryonic ages described in this chapter are those for the mouse. Approximately

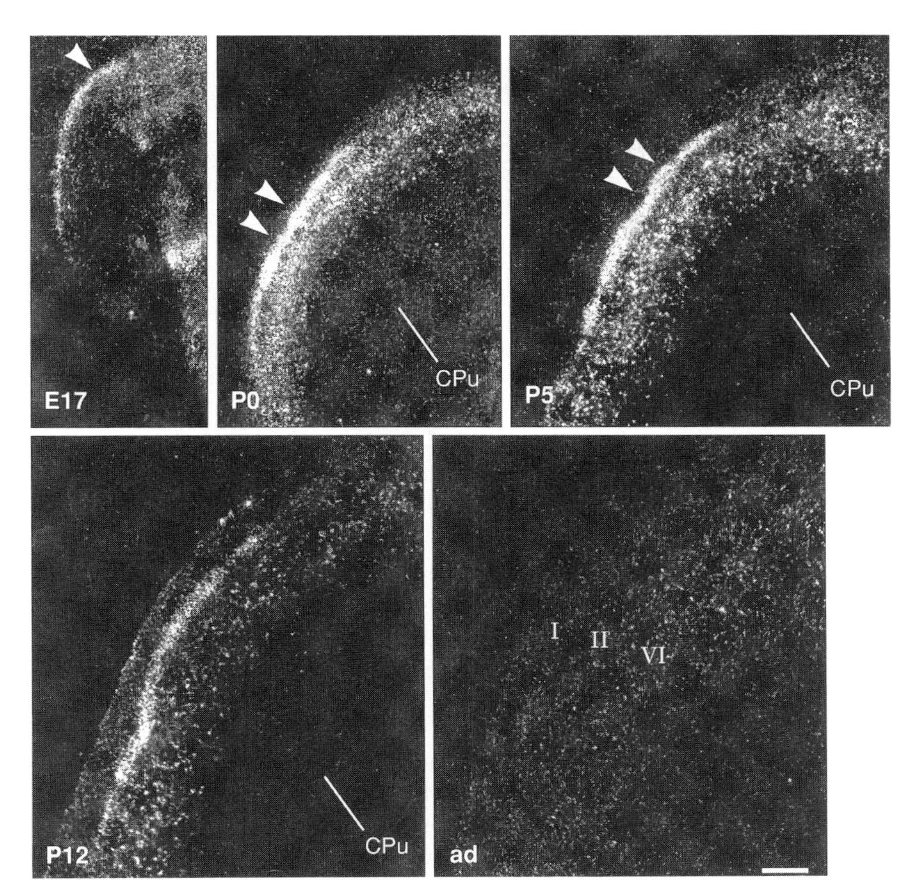

FIG. 3.2. Enlargement of X-ray film autoradiographs illustrating the expression of a kainate receptor subunit, GluR5, in the rat sensory cortex at different stages of brain development (E17, P0, P5, P12; ad, adult). White arrowheads mark an intense line of expression in layer II/III cells of the somatosensory cortex. This expression is absent in the adult. CPu, caudate-putamen. (Reproduced from Bahn *et al.*, 1994).

equivalent stages of rat development are given in Table 3.1. For the mouse, all 26 Theiler stages (E0–E17.5) are available on the Internet at http://genex.hgu.mrc.ac.uk/ (Davidson *et al.*, 2001); this site explains how the Theiler stages are defined. The Mouse Atlas CD has the same information (Baldock *et al.*, 2001).

TABLE 3.1
AN APPROXIMATE CORRELATION BETWEEN MOUSE AND RAT EMBRYONIC DEVELOPMENT COMPILED FROM DATA BY GOEDBLOED AND SMITS-VAN PROOIJE (1986), THEILER (1989), RUGH (1991) AND KAUFMAN (1992).

Approximate number of somites	Stage of embryonic development (gestational age in days)	
	Mouse	Rat
1–7	8	9
8–12	8.5	9.8
13–21	9	10.5
22–26	9.5	11.2
27–33	10	12
34–38	10.5	12.5
39–43	11	13

3.2.1 EMBRYOS

To obtain precisely staged embryos the mating period should be as short as possible; only females with a definite postcoital vaginal plug should be used (see Theiler, 1989, or Rugh, 1991, for more details). For greatest efficiency, matings should be set up in the early morning (e.g. about 8 a.m.) for a period of 1 h. Since oestrus normally begins around midnight, morning matings have the advantage that all the receptive eggs are already present in the oviduct and can be fertilized immediately and approximately simultaneously. Such matings reduce ageing errors of embryos to under 2 h. The vaginal plug is visible for only a few hours and from the point of its detection we assign the age embryonic day 0 (E0). With evening matings, the spermatoza must survive in the female reproductive tract until the following ovulation, when ova are ready for fertilization in a sequential manner. Long overnight matings can therefore result in an age range of up to 16 h, with resultant embryos differing in their developmental stages (Juurlink and Federoff, 1980).

If a more exact staging of development is required, individual embryonic ages can be confirmed by:

1. examination of somites (Rugh, 1991)
2. examination of paw, eye or ear development (Rugh, 1991)
3. measurement of crown–rump length (Dunnett and Björklund, 1992).

In addition, E8 and E9 embryos are distinguishable by posture; an E8 embryo is dorsally curved, whereas at E9 the tail faces the ventral surface of the head (Theiler, 1989; Kaufmann, 1992).

3.2.2 Postnatal Animals

For postnatal brains, the first 24 h after birth is termed by us postnatal day 0 (P0), following the reasons outlined in Paxinos *et al.* (1994).

3.3 Tissue Excision

3.3.1 Embryos

We give a brief description on uterine dissection and embryo excision. More detail can be found in Hogan *et al.* (1994), Dunnett and Björklund (1992) and Beddington (1987). The pregnant females are killed by cervical dislocation; the uteri are dissected from the abdomen, then rinsed in ice-cold buffer (200 mM HEPES, pH 7.4) and placed in a second Petri dish containing the same buffer. During dissection, the buffer must be kept cold. We recommend HEPES buffer rather than phosphate-buffered saline as it keeps the tissue mass in better condition. Ideally, the HEPES buffer should be fresh, but can be stored at 4°C for up to 4 days.

For early (E8–E10) embryos there are several methods of preparation.

Uterus

The fastest and simplest method is to leave the embryos *in utero*. The uterus is cut into segments of one to three decidua each and arranged on an aluminium foil strip. The foil is placed on dry ice until the tissue completely freezes (approx. 5 min). This method is useful for E10 embryos, but tissue morphology is not ideally preserved.

Isolating the deciduum

The excised uterus can be portioned into a vascularized dorsal face and a ventral face; the ventral face contains the E8–E10 embryo. Using a dissecting microscope, the decidua (egg-shaped structures inside the thick muscular myometrium) are dissected by sliding very fine forceps (Dumont no. 5) between the muscle layer and the deciduum on the dorsal side and peeling away the myometrium. Care should be taken not to harm the deciduum. This is pressurized and may rupture, expelling and damaging the embryos. At these early stages the tissue integrity is still loose. It is advisable to fix either the whole deciduum or the completely excised embryo. The embryo is transferred into fixative with a small spatula or the embryo isolated (see below).

Embryos

The isolated deciduum has ventral (usually narrower and lighter) and dorsal (larger and darker) ends. The unfixed deciduum is placed into fresh buffer and the middle pinched with very fine forceps to separate the ventral and dorsal ends. The E8–E10 embryo in the ventral end can be obtained by gently pulling apart the ventral deciduum. The embryo should still be surrounded by the yolk sac and amnion. These membranes are removed by tearing them with forceps. The embryo is then transferred into fixative with an autoclaved toothpick or sterile Pasteur pipette.

Older embryos (E11–E20) can be processed like the postnatal brains. They are excised directly from the uterus by cutting open the myometrium and popping out an embryo within its yolk sac. The yolk sac can be either torn or cut open. After rinsing the embryo, it is placed on its side on aluminium foil and excess buffer carefully removed with the edge of a tissue or the tip of a triangular piece of filter paper.

The aluminium foil with embryo is then placed on dry ice until the embryo is frozen. They are stored wrapped in parafilm at –70°C for up to six months.

Prefixing and embedding E8–E10 tissue.

For early stages, prefixation and sucrose infiltration produce better tissue preservation and make sectioning easier. Later stages do not need to be prefixed as the tissue integrity is tighter; they can be manipulated as described in Chapter 2.

The E8–E10 decidua or embryos are transferred to a 15-ml Falcon tube containing 10 ml fixative for 5 min (4% PFA, Protocol 1.2, step 2, Chapter 1); then transferred to another tube of fixative for a further 25 min. After this fixation, the tissues are transferred to PBS for 1 min, and then to sucrose (0.5 M in PBS, pH 7.4) until the embryos sink to the bottom (approx. 30 min). After this a further rinse is carried out in PBS.

The decidua, after being frozen on dry ice, can be embedded and sectioned. Fixed E8–E10 embryos must be embedded unfrozen. An embedding chamber (e.g. Peel-a-Way disposable embedding moulds; Polysciences) is one-third filled with embedding fluid (e.g. Tissue-Tek, OCT Compound; Miles). Using a toothpick, several sucrose-infiltrated E8–E10 decidua are placed on the fluid surface. Controlling their orientation is very difficult as the embryos are very small and tend to break in the viscous fluid. More embedding fluid is added on top until the chamber is half-full. The position and depth of the embryos are noted and the chamber placed into dry ice until the embedding fluid completely solidifies. Such embedded embryos can be stored frozen at –70°C for up to six months if wrapped in parafilm. Before sectioning, the plastic chamber is cut away with a razor blade. Then the tissue block is attached to a cryostat specimen holder (Chapter 2). Sections are cut beginning from the top surface.

3.3.2 Postnatal Brains

In principle, the techniques for removing early postnatal brains are the same as for adult brains. After decapitation, the head is cooled on wet ice. At this age, the skull sutures can easily be cut with small scissors. The rear skull plates are removed first, and then a cut made forwards along the sagittal suture. The young brain is soft and fragile, so the scissor points should be advanced only a few millimetres each snip, keeping the interior blade of the scissors horizontal to the brain surface. After the frontal skull plate is cut, the skull plates are folded back with forceps. If the meninges do not tear off with the bone, they are carefully removed from the brain surface with forceps and fine scissors. If the meninges are still present when the brain is removed, they will cut through the tissue like cheese wire. The brain is carefully scooped out with a narrow spatula (5 mm), beginning at the front and working back, allowing the brain to drop gently onto aluminium foil. The brain is frozen on dry ice (Chapter 2).

3.4. Sectioning Tips

Cryostat sectioning and troubleshooting is described in Chapter 2. As for adult tissue, all embryonic and postnatal tissues should be equilibrated to the cutting temperature (–20°C) for 1 h before sectioning. If you do not do this, section compression, expansion or fracturing occurs (see Chapter 2). While cutting, occasional sections should be stained to identify structures and confirm orientation (Fig. 3.1). The air-dried section is dipped in thionin stain (Chapter 1, Protocol 1.9, step 14) for a few seconds, then rinsed in water. After cutting, the sections are fixed and dehydrated as described in Chapter 1, Protocol 1.2.

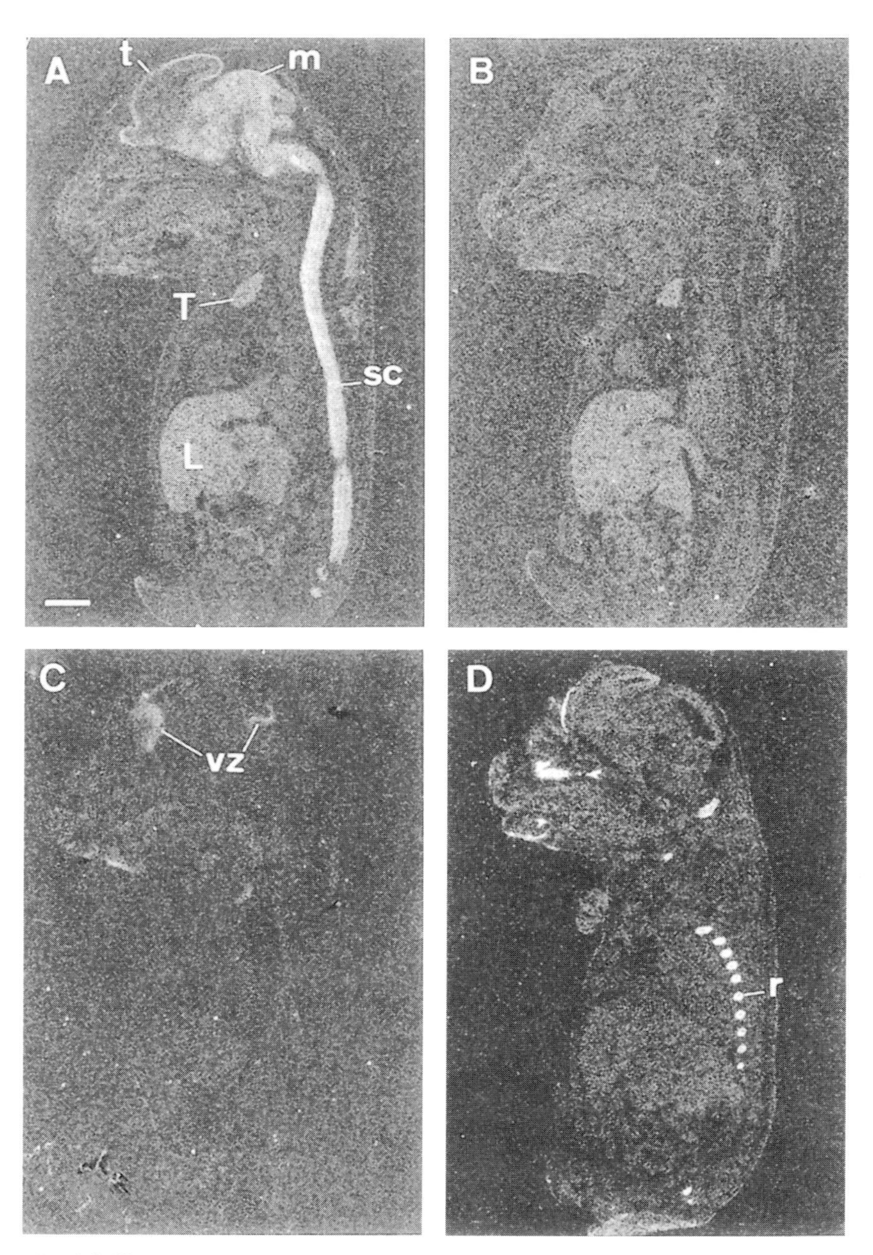

FIG. 3.3. Illustration of non-specific hybridization problems on rat E19 embryo sections. (A) and (B) Total and non-specific hybridization (competition) signals using an NMDA NR1-2 subunit probe. (C) Non-specific hybridization of an NMDA receptor NR1-4 subunit probe which had a long poly(dA) tail. (D) Non-specific hybridization of an NMDA ∂1 receptor subunit

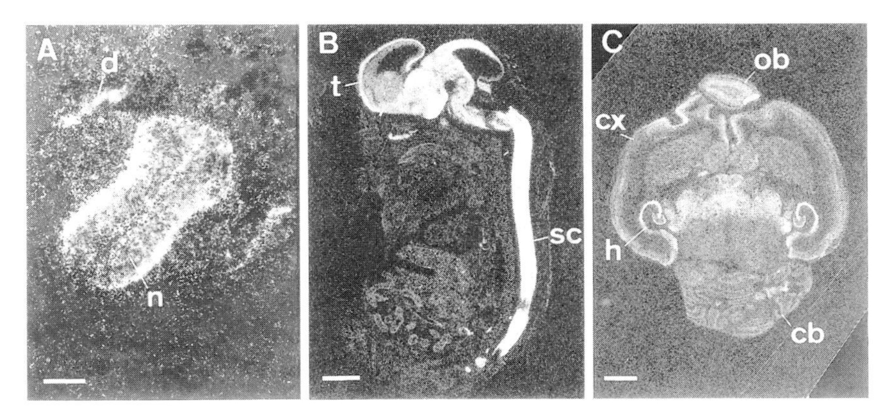

FIG. 3.4. *In situ* hybridization signals. (A) Prefixed mouse E9.5 embryo (Nestin probe, dark-field picture) (P. C. U. Schrotz and W. B. Huttner, unpublished); (B) E17 rat embryo ($GABA_A$ receptor subunit β3 probe – Laurie *et al.*, 1992); (C) rat P6 brain ($GABA_A$ receptor subunit α2 subunit probe – Laurie *et al.*, 1992). d, dorsal root; n, neural tube; others as for Fig. 3.1. Scale bars 0.1 mm (A) and 2 mm (B and C).

Three orientations are possible when sectioning the whole uterus, using the mesometrium as a reference (see Rugh, 1991). When embedding the frozen deciduum, it is orientated with the ventral pole to the top to get transverse sections of the embryo. Sectioning the deciduum vertically gives longitudinal embryo sections. However, excision and freezing may alter the embryo position, so these are only approximate orientations.

Staining of parallel sections to check position and orientation is essential when cutting early embryos (E8–E10), as they are invisible in the white embedding compound (Fig. 3.1A). When cutting coronal or sagittal sections of later embryos (E11–E20), they are orientated on the cutting blockhead downwards (i.e. knife begins cutting section at the head). This minimizes damage to the brain and surrounding tissues (Fig. 3.1B).

FIG. 3.3 *continued*

probe that had a ^{33}P-labelled poly(dA) tail. (Probes were as used in Lomeli *et al.*, 1993 and Laurie and Seeburg, 1994.) L, liver; m, mesencephalon; r, bone, cartilage; sc, spinal cord; t, telencephalon; T, thymus; vz, ventricular zone. Scale bar 2 mm.

3.5 Hybridization Tips

All slide-mounted embryo sections should be hybridized as described for adult brain sections (Chapter 1, Protocols 1.5 and 1.6). (Fig. 3.4.) After hybridization, sections can be exposed to X-ray film, which gives a good picture of the overall expression pattern of the gene (Fig. 3.4).

3.5.1 Non-specific Hybridization

Embryos tend to give a higher hybridization background than adult sections. It can help to acetic anhydride pretreat the section prior to hybridization (Chapter 1, Protocol 1.10).

Peripheral tissues

Non-specific hybridization often occurs on peripheral embryonic tissues (Fig. 3.3A,B). Different oligonucleotide probes show different degrees of this; thus it is essential to run controls in which unlabelled probe is added to the hybridization buffer (see Chapter 1, Protocol 1.5, footnote *a*).

Ventricular zones can trap the probe

Proliferative, cell-dense ventricular zones of the embryonic central nervous system have the same problems as cerebellar granule cells: they non-specifically trap probes, especially if their poly(dA) tails are too long (Fig. 3.3C) (see Chapter 1, Section 1.7.6).

Avoid sections drying out

If the section slightly dries out during hybridization or washing, this causes a strong increase in background (see Chapter 1, Fig. 1.18E). Avoid this by spreading hybridization buffer evenly on the slide; and ensure no overhang of the parafilm coverslip. For the wash step, transfer each slide individually into a rack submerged in 1 × SSC buffer.

Avoid [^{33}P]dATP for embryo sections

Hybridizing ^{33}P-labelled oligonucleotides to embryonic sections is not recommended. This can sometimes give non-specific labelling of cartilaginous tissue (Fig. 3.3D) However ^{33}P can be used for postnatal brain sections.

3.5.2 Avoid Excessive Staining after Developing Emulsions

Hybridized embryo and early postnatal sections can be dipped in photographic emulsion and stained exactly as described in Chapter 1, Protocol 1.9. However, brief (30 s) stain immersion times should be used. Embryonic tissue binds the dye faster than adult tissue. If the section is too heavily stained, it will be hard to see the silver grains (Fig. 3.3A).

3.6 Structure Identification

The developing mouse brain has been thoroughly mapped (Theiler, 1989; Rugh, 1991; Kaufman, 1992; Jacobowitz, 1998; Kaufman and Bard, 1999; Williams, 2000). There is a valuable Internet site: the Mouse Atlas Database, which contains 3D models of mouse embryos and many useful applications (Baldock *et al.,* 2001; Davidson *et al.,* 2001; http://genex.hgu.mrc.ac.uk/). For rat brain development use Paxinos *et al.* (1994) and/or Altman and Bayer (1995). Structures of the postnatal rodent brain can usually be identified by comparison with an adult brain atlas (Paxinos and Watson, 1998; Rosen *et al.,* 2000; Paxinos and Franklin, 2001).

Acknowledgements

These procedures were originally established in the laboratories of Professors P.H. Seeburg and W.B. Huttner.

References

Altman, J. and Bayer, S. A. (1995). "Atlas of Prenatal Rat Brain Development". CRC Press, Boca Raton.

Bahn, S., Volk, B., and Wisden, W. (1994). *J. Neurosci.* **14,** 5525–5547.

Baldock, R., Bard, J., Brune, R., Hill, B., Kaufman, M., Opstad, K., Smith, D., Stark, M., Waterhouse, A., Yang, Y., and Davidson, D. (2001). *Brief Bioinform.* **2,** 159–169.

Beddington, R. (1987). *In* "Mammalian Development: a Practical Approach". (M. Monk, Ed.), pp.43—69. IRL Press, Oxford.

Burazin, T. C. and Gundlach, A. L. (1999). *Brain Res. Mol. Brain Res.* **73,** 151–171.
Burazin, T. C., Larm, J. A., Ryan, M. C., and Gundlach, A. L. (2000). *Eur. J. Neurosci.* **12,** 2901–2917.
Davidson, D. and Baldock, R. (2001). *Nature Rev. Genet.* **2,** 409–417.
Davidson, D., Bard, J., Kufmann, M., and Baldock, R. (2001). *Trends Genet.* **17,** 49–51.
Dunnett, S. and Björklund, A. (1992). *In* "Neural Transplantation – a Practical Approach". (S. B. Dunnett and A. Björklund, Eds) pp.1–19, IRL Press, Oxford.
Giffard, R. G., Papadopoulos, M. C., van Hooft, J. A., Xu, L., Giuffrida, R., and Monyer, H. (2000). *J. Neurosci.* **20,** 1001–1008.
Goedbloed, J. F. and Smits-van Prooije, A. E. (1986). *Acta Anat* **125,** 76–82
Greene, W. K., Bahn, S., Masson, N., and Rabbitts, T. H. (1998). *Mol. Cell Biol.* **18,** 7030–7037.
Hogan, B., Beddington, R., Constantini, F., and Lacy, E. (1994). "Manipulating the Mouse Embryo". Cold Spring Harbor Laboratory Press, Cold Spring Harbor, NY.
Hormuzdi, S. G., Pais, I., LeBeau, F. E., Towers, S. K., Rozov, A., Buhl, E. H., Whittington, M. A., and Monyer, H. (2001). *Neuron* **31,** 487–495.
Jacobowitz, D. M. (1998). "Chemoarchitectonic Atlas of the Developing Mouse Brain. CRC Press, Boca Raton.
Juurlink, B. H. and Federoff, S. (1980). *In Vitro* **15,** 86.
Kaufman, M. H. (1992). "The Atlas of Mouse Development", revised edition. Academic Press, London.
Kaufman, M. H. and Bard, J. (1999). "The Anatomical Basis of Mouse Development". Academic Press, London.
Laurie, D. J. and Seeburg, P. H. (1994). *J. Neurosci.* **14,** 3180–3194.
Laurie, D. J., Wisden, W., and Seeburg, P. H. (1992). *J. Neurosci.* **12,** 4151–4172.
Livesey, F. J. and Hunt, S. P. (1997). *Mol. Cell Neurosci.* **8,** 417–429.
Lomeli, H., Sprengel, R., Laurie, D. J., Köhr, G., Herb, A., Seeburg, P. H., and Wisden, W. (1993). *FEBS Lett.* **315,** 318–322.
Monyer, H., Seeburg, P. H., and Wisden, W. (1991). *Neuron* **6,** 799–810.
Monyer, H., Burnashev, N., Laurie, D. J., Sakmann, B., and Seeburg, P. H. (1994). *Neuron* **12,** 529–540.
Nase, G., Weishaupt, J., Stern, P., Singer, W., and Monyer, H. (1999). *Eur. J. Neurosci.* **11,** 4320–4326.
Paxinos, G. and Franklin, K. (2001). "The Mouse Brain in Stereotaxic Coordinates", second edition. Academic Press, London.
Paxinos, G. and Watson, C. (1998). "The Rat Brain in Stereotaxic Coordinates", fourth edition. Academic Press, London.
Paxinos, G., Ashwell, K. W. S., and Törk, I. (1994). "Atlas of the Developing Rat Nervous System". Academic Press, London.
Ringwald, M., Baldock, R., Bard, J., Kaufman, M., Eppig, J. T., Richardson, J. E., Nadeau, J. H., and Davidson, D. (1994). *Science* **265,** 2033–2034.
Rosen, G. D., Williams, A. G., Capra, J. A., Connolly, M. T., Cruz, B., Lu, L., Airey, D. C., Kulkarni, K., and Williams, R. W. (2000). *Int. Mouse Genome Conference* **14,** 166. www.mbl.org.
Ryan, M. C., Loiacono, R. E., and Gundlach, A. L. (1997). *Neuroscience* **78,** 1113–1127.
Rugh, R. (1991) "The Mouse. Its Reproduction and Development". Oxford University Press, Oxford.
Sinclair, A. M., Gottgens, B., Barton, L. M., Stanley, M. L., Pardanaud, L., Klaine, M., Gering, M., Bahn, S., Sanchez, M., Bench, A. J., Fordham, J. L., Bockamp, E., and Green, A. R. (1999). *Dev. Biol.* **209,** 128–142.
Theiler, K. (1989). "The House Mouse. Atlas of Embryonic Development". Springer-Verlag, Heidelberg.

Voigt, M. M., Laurie, D. J., Seeburg, P. H., and Bach, A. (1991). *EMBO J.* **10,** 4017–4023.
Williams, R. W. (2000). *In* "Mouse Brain Development" (A. F. Goffinet and P. Rakic, Eds), pp. 21–49, Springer, New York.
Wisden, W., Parker, E. M., Mahle, C. D., Grisel, D. A., Nowak, H. P., Yocca, F. D., Felder, C. C., Seeburg, P. H., and Voigt, M. M. (1993). *FEBS Lett.* **333,** 25–31.

Chapter 4

PROCESSING RETINAL TISSUE FOR *IN SITU* HYBRIDIZATION

F. Müller

Institut für Biologische Informationsverarbeitung 1, Forschungszentrum Jülich, D-52425 Jülich, Germany

4.1 Introduction

This chapter is for non-retinologists who need to perform *in situ* hybridizations on the retina. The vertebrate retina originates from the diencephalon, and is thus a part of the central nervous system. It is a well-organized, layered structure, which makes identification of cell types relatively easy (Fig. 4.1). The retina contains most of the neurotransmitters, neuropeptides and their respective receptors found in other brain regions, but is more easily approachable for anatomical and physiological experiments: it is a "natural brain slice". For reviews on the retina see Masland (2001) or the Webvision Homepage (http://webvision.med.utah.edu). The expression and localization of different receptor classes has been successfully studied using *in situ* hybridization with the method described earlier (Müller *et al.*, 1992; Brandstätter *et al.*, 1994; Greferath *et al.*, 1995; Hartveit *et al.*, 1995; reviewed in Wisden *et al.*, 2000 – see Fig. 4.1 for ionotropic glutamate receptors). The retinal *in situ* hybridization protocol does not differ from the standardized procedure described in Chapter 1 and will not be discussed in detail here, except to mention that it is best to pretreat the sections with acetic anhydride and chloroform (see Chapter 1, Protocol 1.10 and Section 4.4 of this chapter). Instead, I focus on the dissection and preparation of the retina for the *in situ* hybridization technique.

INTERNATIONAL REVIEW OF NEUROBIOLOGY, VOL. 47
ISSN/02 $00.00

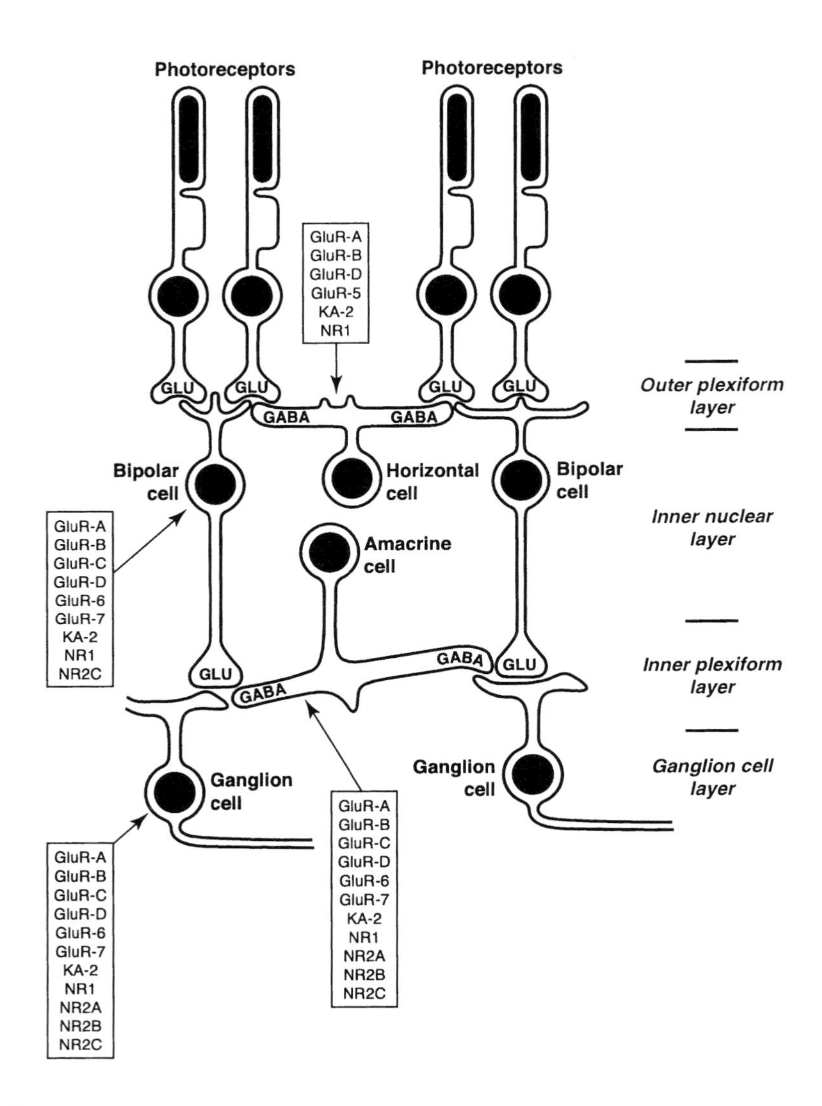

FIG. 4.1. The cell types in the rat retina and their expression of the AMPA, NMDA and kainate receptor subunit mRNAs summarized from *in situ* hybridization data (circuit diagram adapted from Barnstable, 1993). (Reproduced from Wisden *et al.*, 2000.)

I first briefly describe the anatomy of the retina to cover the most frequently asked questions and to facilitate the interpretation of *in situ* hybridization data (Section 4.2), and then describe how to prepare retinal tissue for *in situ* hybridization (Sections 4.3 and 4.4).

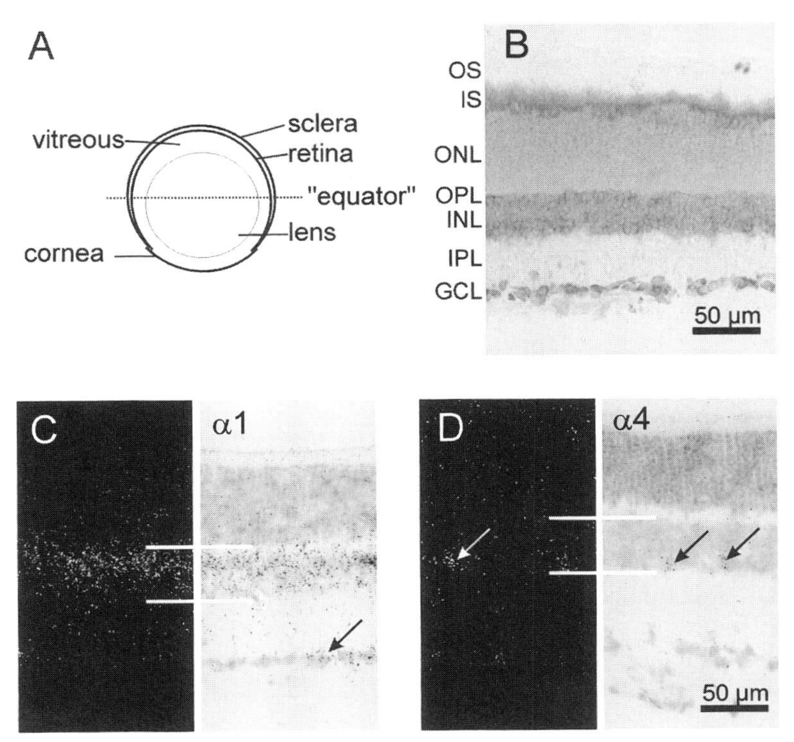

FIG. 4.2. (A) Scheme of a rodent eye. (B) Nissl staining of a vertical section through the rat retina. OS, outer segments; IS, inner segments; ONL, outer nuclear layer; OPL, outer plexiform layer; INL, inner nuclear layer; IPL, inner plexiform layer; GCL, ganglion cell layer. (C) *In situ* hybridization with α ^{35}S-labelled oligonucleotide specific for the α1 subunit of $GABA_A$ receptors in dark-field (left) and bright-field (right, counterstained). Silver grains are predominantly found over the outer half of the INL, indicating expression in bipolar cells and in a few large cells in the GCL, probably ganglion cells. White bars mark borders of the INL. (D) *In situ* hybridization with α ^{35}S-labelled oligonucleotide specific for α4 subunit of the $GABA_A$ receptor. Conditions as in C. Only a few somata in the inner half of the INL are labelled (arrows), indicating expression in a small subset of amacrine cells.

4.2. Anatomy of the Retina

The vertebrate retina is a well-layered neuronal tissue, in most species 150–200 μm in thickness. Figure 4.2B shows a vertical section (perpendicular to the surface) of the retina. The layers closer to the optical apparatus and hence to the incoming light are called "inner" or "proximal" retina; those further away are "outer" or "distal" retina. The somata of the photoreceptors, the rods and cones, are found in the outer nuclear layer (ONL). They grow elongated processes, the inner segments (IS) that are the site of protein synthesis. Therefore, probes for photoreceptor mRNA will mainly label the IS layer, with weaker label over the ONL. The outer segments (OS) contain all proteins necessary for photoelectrical transduction, but no mRNA. In most species, only a small percentage of all photoreceptors are cones (in rodents 1–3%). In the outer plexiform layer (OPL), which is usually very thin in rodents, the photoreceptors are hooked onto the second-order neurons (bipolar and horizontal cells), whose somata are found in the outer half of the inner nuclear layer (INL) (see Fig. 4.1). The bipolar cells convey their information to ganglion cells and amacrine cells in the inner plexiform layer (IPL) (Fig. 4.1). Ganglion cell somata are found in the ganglion cell layer (GCL); their axons leave the eye at the optic disc (blind spot).

In addition to this vertical pathway, lateral connections and feedback synapses are made by horizontal cells in the OPL and by amacrine cells in the IPL. Amacrine cell somata are found in the inner half of the INL and as "displaced" amacrine cells in the GCL. Therefore, small somata in the GCL cannot automatically be classified as ganglion cells. Each of the retinal cell classes can be divided into multiple subpopulations. In most species there is one type of rod and two to three types of cone photoreceptors, two types of horizontal cells, 10 types of bipolar cells, around 30 types of amacrine cells and 10–20 classes of ganglion cells. Thus, whereas it is relatively easy to distinguish bipolar cells from amacrine cells based on the location of their somata, the differentiation between different amacrine cell classes requires much more sophisticated methods. Figures 4.2C and D show two examples of *in situ* hybridizations performed with ^{35}S-labelled oligonucleotides specific for the mRNA of the $\alpha 1$ and $\alpha 4$ subunit genes of $GABA_A$ receptors, respectively. For $\alpha 1$, the signal was most prominent throughout the outer half of the INL (the borders of the INL are marked by white bars), i.e. in bipolar cells and in some large cells in the GCL, probably ganglion cells. In contrast, $\alpha 4$ mRNA was only found in a few somata in the inner half of the INL, i.e. in a subpopulation of amacrine cells. Glia cells are also found in the retina. Microglial cells are distributed throughout the inner retina; Müller cells have their somata in the exact middle of the INL and extend their processes vertically throughout the retina; astrocytes are found in the GCL and the nerve fibre layer.

In most species, an area centralis with higher cell density has developed. In rodent retinas, which are mostly used for *in situ* hybridization studies, the density gradient is only small and the retina can almost be considered as homogeneous. A specialized fovea, in which the inner retinal layers have been shifted aside as in humans, does not exist. A special feature of the mouse retina is the non-uniform distribution of cone photoreceptors. In the upper part of the retina the cones are mainly red–green sensitive, whereas in the lower part the cones are mainly UV–blue sensitive. For photoreceptor-specific probes in mouse, therefore, it may be a good idea to section the retina along the vertical axis.

In principle, the retina can be sectioned within the eyecup. In this case, however, only those sections taken close to the optical axis of the eye will be vertical. Sections taken further away will be cut more or less oblique. This will change the appearance of the layering and identification of the cells will be more difficult. This effect is more pronounced in small eyeballs, e.g. in mouse. Also, the retina will be strongly curved. If only few sections are needed, or if you have little experience with the handling of delicate structures such as the retina, you may find it easier to section the retina within the eyecup. However, it is usually better to remove the retina from the eyecup and mount it such that the whole retina can be sectioned vertically. It is possible to section the retina in the complete unfixed eyeball; however, the retina often detaches and folds or the large lens becomes dislocated during sectioning, which may damage the retina. We have found that tissue preservation is much better if the retina is fixed before sectioning.

4.3 Preparation of the Retina

4.3.1. Preparation of the Eye

Experimental animals are deeply anaesthetized with isoflurane or other convenient narcotics and decapitated. Putting one finger onto the skin above the eye and one below and moving them apart, opens the eye, allowing the eyeball to come out of the orbita. With the eye in this position, the eyeball is dissected free with small scissors (optimally with curved blades) using small cuts and taking care not to cut into the posterior eyeball. The eyeball is transferred into a small Petri dish with 4% paraformaldehyde (PFA Chapter 1, Protocol 1.2, Step 2) under a binocular dissecting microscope. The eye must be completely submersed. Larger pieces of conjunctiva and muscle can be cut away, but it is important to leave enough so that the eye can be gripped with fine-tipped Dumont forceps. The eye is held firmly at the conjunctiva and a small incision made in the eyeball at the equator with a scalpel blade or an injection needle. This incision is then cut further open with small dissecting scissors (optimally spring scissors, blades approx. 5 mm long). The aim is to open the eye with an encircling cut around the equator (see line in Fig. 4.2A), such that the anterior part with the cornea and lens can be removed and the posterior part with the retina is exposed to the fixative. It is best to cut the eye open in small steps and to hold the scissors parallel to the surface of the eye, in order not to hit the lens. The eye is held with Dumont forceps close to the spot where the cut is made so that it cannot move. It is important to keep the eye submersed. When the eye is opened completely, the anterior part is carefully removed. The lens, which is very large in rodents, will come off with the cornea. The vitreous is a very sticky fluid that covers the retina. Experienced researchers can try to remove the vitreous by holding the eyecup with one Dumont forceps at the sclera, gripping the vitreous with a second forceps above the retina and pulling it out. Care is needed, as the vitreous is rather sticky and it may lift the retina. In this case, it is better to leave the vitreous within the eyecup. The eyecup is left with the retina in the fixative for 30–60 min. After fixation, the eyecup is rinsed several times with phosphate buffer (PB) or PBS, then transferred to 10% sucrose/PB for 30 min and finally to 30% sucrose/PB for 1– 2 h for cryoprotection.

4.3.2 Sectioning of the Eyecup

The eyecup is transferred from the sucrose solution into a large drop of Tissue-Tek (Miles Scientific). It is important to ensure that the sucrose solution within the lumen of the eyecup is exchanged by Tissue-Tek. The eyecup is left in the Tissue-Tek for a minute and a slide or a plastic Petri dish is covered with a piece of parafilm. Next the eyecup is transferred with the opening downwards into a drop of fresh Tissue-Tek on the parafilm, covered with sufficient Tissue-Tek and frozen in the cryostat. After freezing, the block is removed from the parafilm, turned upright by 90 degrees and mounted in some Tissue-Tek on a cryostat chuck. As the block is freezing, the Tissue-Tek is spread with a scalpel blade from the chuck to the top of the block to improve adhesion of the block. The block can now be sectioned. The first sections will be very oblique and should be discarded. While cutting into the block, some sections should be stained (see below) to check the tissue quality and the orientation of the plane of section. The ganglion cell layer in rat is usually only one to two rows of cells; however, if cut tangentially, it appears multilayered. It is therefore a good sign to start collecting sections if the ganglion cell layers are one to two rows of cells thick. From time to time, sections should be stained to check for changes in the tissue.

4.3.3 Sectioning of the Retina

The retina is dissected from the eyecup at the end of the 30% sucrose step. All steps are carried out with the eyecup submersed in the sucrose solution. By rolling over with a round scalpel blade, peripheral pieces of tissue (both retina and sclera) are cut away from four sides of the eyecup. This yields a square-shaped piece of tissue (in mouse approx. 3 mm wide, in rat 5–6 mm). It is now easy to isolate the square of retina from the rest of the eyecup by cutting axons and blood vessels below the optic disc with small scissors. The retina must now be mounted vertically. With a fine brush, the retinal piece is transferred to a drop of sucrose on a piece of parafilm that covers a slide or the lid of a small plastic Petri dish. If the retina rolls strongly upwards (towards the GCL), it can be mounted GCL-down flat onto the parafilm with two fine brushes. Very gently holding the retina flat with a brush, the sucrose solution is aspirated with small pieces of filter paper. When the retina is completely flat, it is covered with a drop of Tissue-Tek and frozen in the cryostat. The frozen block is then removed from the parafilm and mounted vertically on a cryostat chuck in some Tissue-Tek, with one border of the retina at the top. While the retina is freezing, the Tissue-Tek is spread with a scalpel blade from the chuck to the top of the block for better adhesion of the block. The retina is now ready for sectioning. The retina is cut along the long axis of the tissue.

4.4 Sectioning and *In Situ* Hybridization

Cryostat sectioning is discussed in detail in Chapter 2. We cut 10–20 μm sections at around –20 to –24°C. If the sections roll while cutting, the temperature can be raised slightly (see Chapter 2). The first sections should be discarded. From time to time, some sections should be stained to check tissue quality. Sections are dried briefly to the slide, fixed with 4% paraformaldehyde (see Chapter 1, Protocol 1.2), dipped in phosphate buffer (PB) (Chapter 1, Protocol 1.2), briefly stained with some drops of staining solution (stock solution 1% methylene blue in 1% sodium borate in distilled water; for usage, dilute 1 : 10 with distilled water), dipped in PB, covered with a coverslip and viewed under the microscope.

For hybridization, sections (several per slide) are collected on silanized or on poly-L-lysine-coated slides (Chapter 1, Protocol 1.1) and air-dried for 30 min to 2 h. They are then fixed onto the slides with 4% PFA for 5 min, followed by washing, dehydration and ethanol storage as described in Chapter 1, Protocol 1.2. For ^{35}S-labelled probes an acetylation step may be included (Chapter 1, Protocol 1.10). We do not use proteinase K steps. For details of hybridization, follow Protocol 1.5 described in Chapter 1. With radioactive probes, the resolution of X-ray films is too low for retinal sections. Slides need to be dipped in photographic emulsion (Chapter 1, Protocol 1.9). Exposure times need to be determined empirically. We usually expose retinal sections at least 50% longer than brain sections hybridized in parallel with the same probe. This might reflect the fact that the somata of retinal neurons are smaller than average brain somata and might contain lower amounts of mRNA.

REFERENCES

Barnstable, C. J. (1993). *Curr. Opin. Neurobiol.* **3,** 520–525.

Brandstätter, J. H., Hartveit, E., Sassoe-Pognetto, M., and Wässle, H. (1994). *Eur. J. Neurosci.* **6,** 1100–1112.

Greferath, U., Grünert, U., Fritschy, J.M., Stephenson, A., Möhler, H., and Wässle, H. (1995). *J. Comp. Neurol.* **353,** 553–571.

Hartveit, E., Brandstätter, J. H., Enz, R., and Wässle, H. (1995). *Eur. J. Neurosci.* **7,** 1472–1483.

Masland, R. H. (2001). *Nature Neurosci.* **4,** 877–886.

Müller, F., Greferath, U., Wässle, H., Wisden, W., and Seeburg, P. (1992). *Neurosci. Lett.* **138,** 179–182.

Wisden, W., Seeburg, P. H., and Monyer, H. (2000). *In* "Handbook of Chemical Neuroanatomy, Vol. 18 Glutamate" (O. P. Ottersen and J. Storm-Mathisen, Eds), pp. 99–143, Elsevier, Amsterdam.

CHAPTER 5

PROCESSING THE SPINAL CORD FOR *IN SITU* HYBRIDIZATION WITH RADIOLABELLED OLIGONUCLEOTIDES

A. Berthele and T.R. Tölle

Department of Neurology, Technical University Munich, Moehlstr. 28, 81675 Muenchen Germany.

5.1 Introduction

The spinal cord receives sensory information from trunk and limbs and is the final station for movement commands. Its anatomy and physiology are well defined and detailed knowledge about the segmental spinal cord circuitry makes this structure ideal for experimental studies. As the first station to integrate somatosensory information in the CNS, the spinal cord is of outstanding interest to the field of pain research and therapy, and is extensively studied in models of traumatic injury or regeneration. Here we describe the application of the *in situ* hybridization (ISH) method of Chapter 1 of this book to investigate gene expression in the spinal cord of the adult rodent. We have extensively used this method to investigate the expression of neurotransmitter systems in the spinal cord (e.g. Wisden *et al.*, 1991; Tölle *et al.*, 1993, 1995a, b; Berthele *et al.*, 1999; Palmer *et al.*, 1999) and to characterize neuroplastic changes following noxious sensory input in various experimental pain models (e.g. Wisden *et al.*, 1990; Palmer *et al.*, 1999; Berthele *et al.*, 2000).

ISSN/02 $00.00

5.2 Spinal Cord Preparation

Portions of spinal cord are dissected from non-perfusion-fixed animals. Animals are terminally anaesthetized and decapitated just prior to spinal cord preparation. To expose the spinal cord *in situ,* dorsal laminectomy is performed according to standard procedures and using conventional surgical instruments such as small scissors, dissecting forceps and small bone rongeurs. The extent of laminectomy necessary depends on the spinal cord region of interest. For example, to dissect the lumbar spinal cord (L_3 to S_1) laminectomy theoretically has to be performed from T_{10} to S_1, but it has to be considered that the lumbar spinal segments are displaced relative to the vertebrae in this region. Therefore for practical purposes, it is much easier to orientate the laminectomy by the level of the last rib and to extend it to about 2–3 cm. Care has to be taken to avoid damage to the dura and the spinal cord tissue underneath. Portions of spinal cord are then further identified by anatomical landmarks, mainly the lumbar and cervical enlargements, and cut horizontally above and below using a razor blade. To avoid problems during tissue sectioning, the length of each spinal cord portion to be dissected should not exceed 1 cm. The dural sac is opened using small scissors and removed. Dorsal and ventral roots are severed and the spinal cord portion is carefully removed from the spine. After removing any adherent dura and blood, the tissue is placed onto a small piece of aluminium foil and frozen on dry ice. Frozen spinal cord portions are stored at –80°C until sectioning; the mRNA content and quality is stable for years.

The following points are crucial to optimize tissue quality:

- Compared with brain tissue, the spinal cord is rich in RNases. The entire procedure (from decapitation to freezing) should be completed in approx. 5 min to avoid mRNA degradation (see Chapter 1, Section 1.5.1 and Chapter 6, Section 6.2.3 for discussion on mRNA stability in post-mortem tissue).
- To preserve tissue morphology manipulation of the spinal cord portions has to be minimized.
- To ease tissue sectioning, remaining dura, blood and any rinsing solution such as Ringer or artificial CSF have to be carefully removed *before* freezing the tissue.
- In terms of tissue morphology and preservation, freezing the tissue on dry ice, which is quick and gentle, seems to be superior to shock-freezing in liquid nitrogen (see also Chapter 2).

The dissection procedure of spinal cord from rat or mice is basically the same. The tiny mouse spinal cord is, of course, more difficult to handle; but on the other hand, laminectomy is easier.

5.3 Spinal Cord Sectioning

5.3.1 General Methodology

Transverse sections of spinal cord are cut on a cryostat, thaw-mounted onto coated glass slides, postfixed and stored until hybridization.

For sectioning, tissue is removed from the freezer and immediately mounted to the specimen stage. Since spinal cord portions are small and thin in diameter, they thaw very easily. To prevent this the spinal cord is mounted in a "upright" position to the specimen stage with only one end embedded in the freezing medium (see Fig. 5.1C), which additionally should be pre-cooled (see also Chapter 2, Cryostat sectioning).

For standard applications, slice thickness of 12–14 μm cut at about –20°C is optimal concerning histology and mRNA preservation and allows unproblematic routine slicing.

However, once started, the question arises when to start collecting the tissue slices. Macroscopically, the exact position of distinct spinal cord segments within the tissue portion that has been dissected cannot be identified: segments only differ in more or less subtle changes in their lamination patterns; or the proportion of grey to white matter; or the presence of typical groups of motoneurons. To identify distinct spinal cord segments the histology of the slices prepared is needed, and this can be checked definitely only after conventional staining procedures. However, this is time consuming and inconvenient to do during the slicing session.

To obtain a first but sufficient impression, it is instead much easier and useful to observe single slices thawing under a microscope in dark-field illumination. Cell bodies thaw much quicker than neuropilar structures and therefore the histology can be nicely judged for a short moment with no need for further time-consuming staining procedures.

Slices are stuck (by pressing the pre-cooled slide against the slice) and thaw-mounted onto coated glass slides. We recommend poly-L-lysine-coated slides (Chapter 1, Protocol 1.1). Other coating techniques are frequently used, for example Vectabond (Vector Laboratories). The use of positively charged slides may be helpful. Gelatin-subbed slides do not work: slices will come off during the washing steps after hybridization. Moreover, we advise

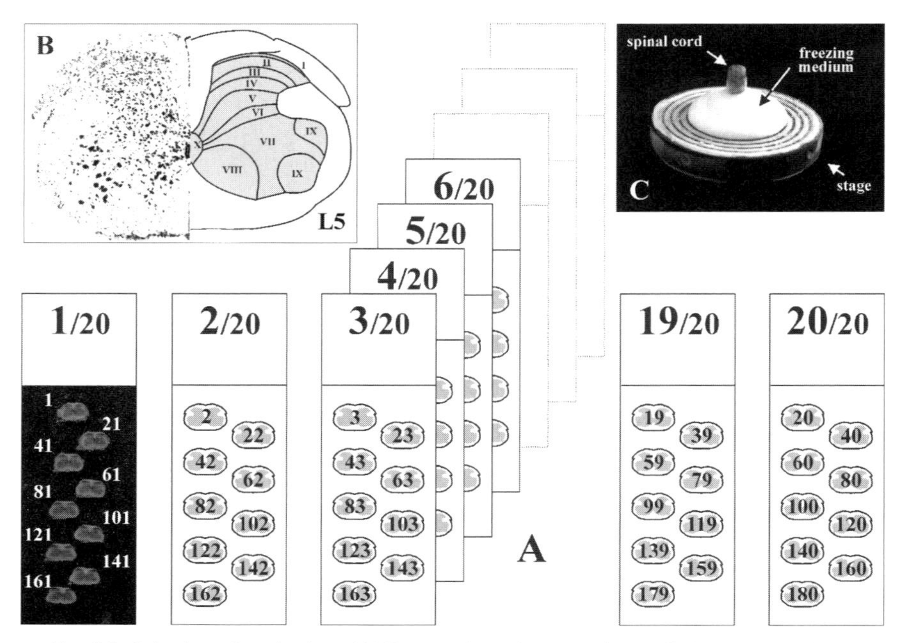

FIG. 5.1. Spinal cord sectioning. (A) Preparation of tissue "aliquots". Consecutive slices are distributed to consecutive slides; see text for details. Given that 20 aliquot slides with nine slices (12 μm) each are prepared, a spinal cord tissue portion of approx. 2.1 mm length is equally represented on each slide. (B) (left) Thionin-stained cell bodies of a slice of adult rat lumbar spinal cord L5 segment. (right) Matching schematic depicting L5 segment lamination. (C) Mounting spinal cord to the microtome stage.

against silane coating since silanization dramatically increases non-specific binding of the oligonucleotide probe to the slide.

Finally, the sections are fixed in 4% paraformaldehyde and stored in 96% ethanol until hybridization (following Protocol 1.2). Storing in ethanol does not harm the ISH signal of the sections even after years. Alternatively, sections can be stored unfixed at –20°C for several weeks, but have to be fixed prior to hybridization. Although there are protocols in the literature which describe ISH with unfixed tissue, we obtained only poor signals with these methods. Using the protocol as described in Chapter 1, ISH may work with perfusion-fixed tissue after tissue pretreatment; however, results are poor.

To assess the histology of stained tissue sections, the papers from Brichta and Grant (1985), Molander *et al.* (1984, 1989) and Nicolopoulos-Stournaras and Iles (1983) are recommended. An example of a thionin-stained lumbar spinal cord section is provided in Fig. 5.1B.

5.3.2 Preparing Tissue "Aliquots"

Owing to metamerity the basic anatomy of the spinal cord does not change from the most cranial/cervical to the most caudal/sacral segments. Of course, afferent inputs as well as the motor circuits do vary depending on the respective segment. Technically, it is difficult to manage preparing spinal cord tissue sections exactly and exclusively from predefined segments. Irrespective of this, to characterize the constitutive mRNA distribution and its changes in experimental paradigms it is often necessary and useful to investigate transcript expression in various segments.

Since transverse spinal cord sections are small in diameter it is possible to collect several slices on each glass slide – not only for practical reasons. Instead of collecting consecutive slices on the same glass slide we propose to distribute consecutive slices to consecutive slides. This method of "aliquoting" provides a number of slides which all represent the same anatomical information of the spinal cord portion sectioned.

For example, to prepare 20 "aliquots" (up to 40 is feasible) 20 pre-cooled and numbered glass slides are stored in the microtome chamber. The first spinal cord section will be stuck to the first glass slide, the second to the second, and so on. Having done this 20 times, the 21st slice will be stuck to the first slide again. In the end, on each respective slide each section will be followed by the forthcoming 20th section. Provided that no slice has been lost during the sectioning procedure, each glass slide will finally bear spinal cord sections covering the whole spinal cord portion which has been cut, shifted by one times the slice thickness from slide to slide (see Fig. 5.1A).

Each slide can now be processed for the detection of different mRNA transcripts. This, for example, allows us to present "side-by-side" comparisons of the differential distribution of these transcripts within the same spinal cord segments as represented on each slide and in the same animal (example given in Fig. 5.3A). For each transcript investigated differences in the segmental expression can be judged directly on each single slide. Moreover, slides with tissue aliquots can be processed with different techniques (e.g. combination of ISH and ligand autoradiography) (see Fig. 5.2A).

5.3.3 Co-expression of mRNAs in the Same Motoneuron

The method of "aliquoting" not only allows for side-by-side comparisons of transcript expression in consecutive sections, it also allows the detection of transcript co-expression in single motoneurons. Since motoneurons are large in diameter, each neuron will be cut two to three times in consecutive sections, at least when the section thickness is reduced to 8–10 μm. The

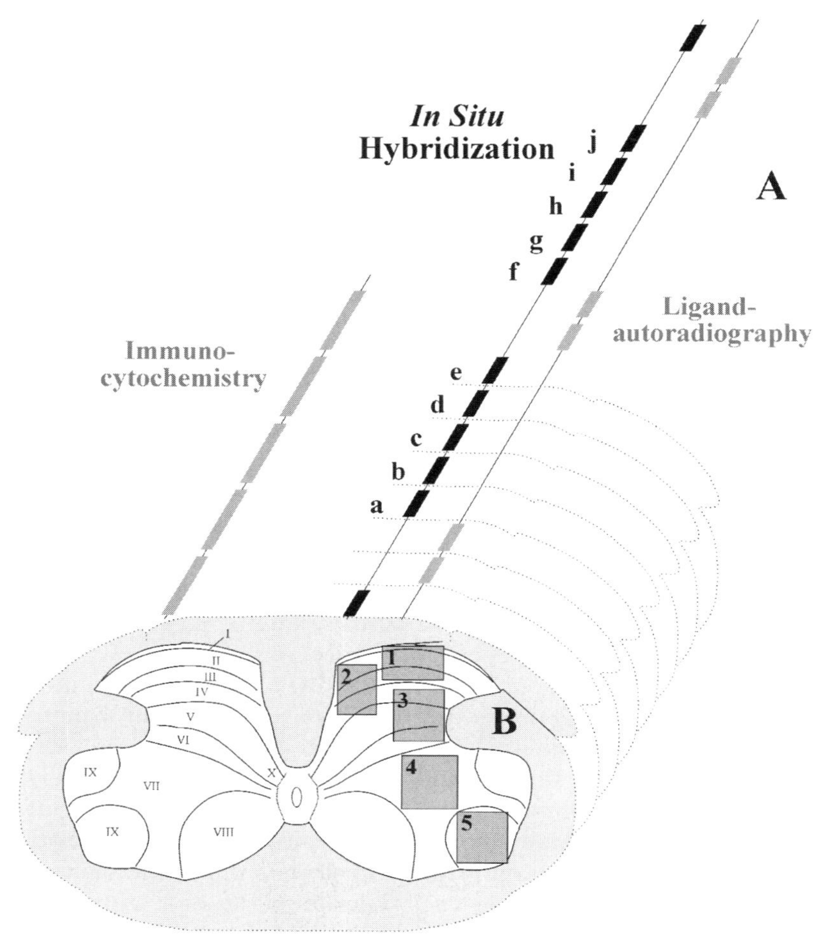

FIG. 5.2. (A) Preparation of tissue "aliquots" is useful in the investigation of the differential distribution of transcripts and proteins and can be applied to post-fixed (e.g. for ISH), unfixed (e.g. for autoradiography) or perfusion-fixed (e.g. for immunocytochemistry) tissue. In ISH, the method allows the side-by-side comparison of various mRNA expression patterns (a to j) in adjacent slices. This information can be combined with data from other histochemical methods on intermingling slides. (B) The regions of interest for densitometry of spinal cord X-ray autoradiographs are depicted in the schematic spinal cord slice.

respective motoneurons can be identified by reconstruction in camera lucida drawings of the consecutive spinal cord slices at low magnification and co-expression can be analysed (see Fig. 5.4 for an example).

A drawback of this method is that even if every motoneuron is sliced about three times, normally only one slice bears nucleus and cytoplasm, where

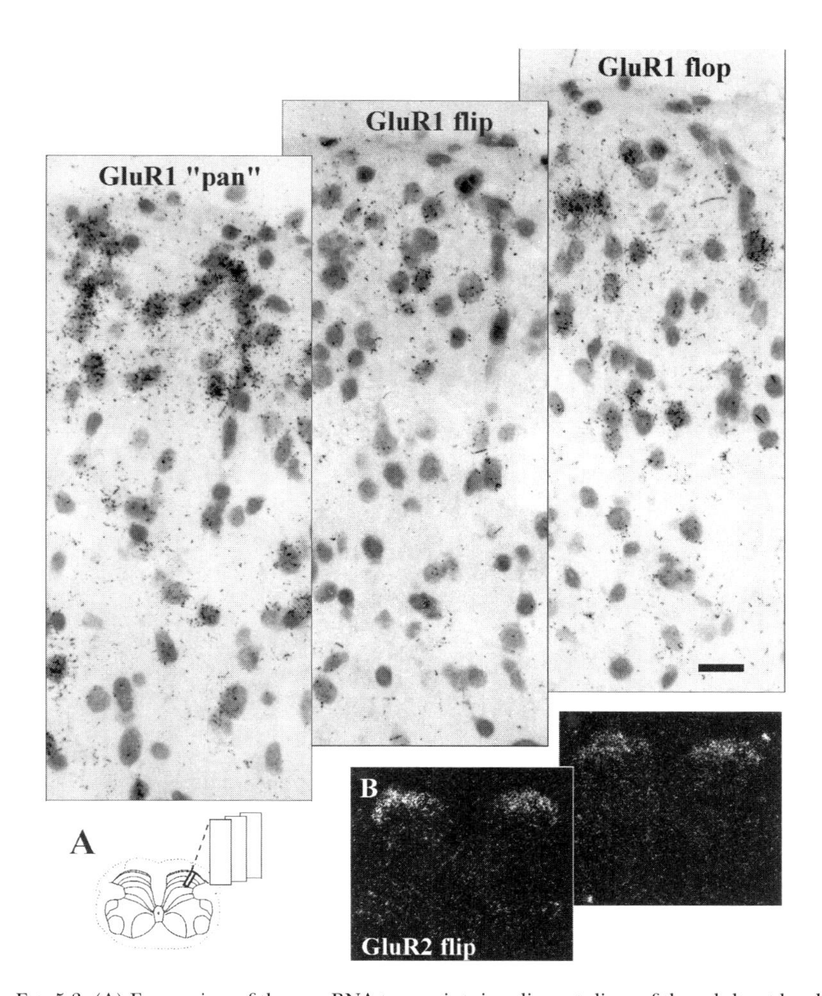

FIG. 5.3. (A) Expression of three mRNA transcripts in adjacent slices of the adult rat lumbar spinal cord dorsal horn. The ionotropic glutamate receptor subunit GluR1 is preferentially expressed in the superficial dorsal horn of the rat lumbar spinal cord. The differential distribution of two GluR1 splice variants (named "flip" and "flop") in comparison to the ISH signal derived from a so-called "pan" probe (which detects both splice variants) is depicted in cellular resolution (see Tölle *et al.*, 1995b, for further information). Clusters of black dots (silver grains) indicate labelled cells. Scale bar 20 μm. (B) X-ray autoradiographs of glutamate receptor subunit GluR2 flip mRNA expression in the adult rat lumbar spinal cord. On the left, the ISH signal is derived from a GluR2 flip complementary radiolabelled 39mer oligonucleotide. On the right, the same oligonucleotide was co-incubated with a second unlabelled oligo in 100-fold excess. This second oligonucleotide, which is complementary to glutamate receptor GluR4 flip, differs only in five nucleotides from the GluR2 flip oligonucleotide (Sommer *et al.*, 1990). However, co-incubation in excess concentration does not compete with GluR2 flip hybridization.

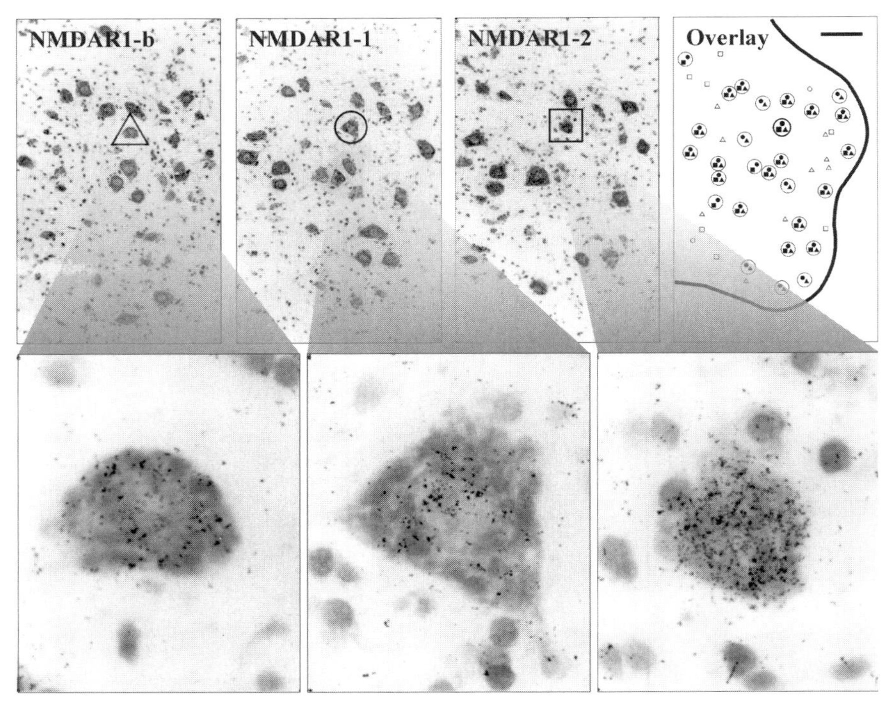

Co-expression of mRNA transcripts in motoneurons. (upper row) Low-power micrographs (thionin stained) of one ventral horn quadrant in three consecutive slices of adult rat lumbar spinal cord. The symbols in the schematic drawing (right) show the anatomical reconstruction of single neurons which could be detected in up to all three consecutive sections. Slices were processed to detect the mRNA expression of splice variants of the glutamate receptor NMDAR1 subunit (triangles, NMDAR1-b; circles, NMDAR1-1; squares, NMDAR1-2) (Laurie and Seeburg, 1994). Scale bar: 100 μm (lower row) High power micrographs depicting the co-localization of the three splice variant transcripts in one single motoneuron (see Tölle et al., 1995a, for further information).

mRNA expression can be judged most reliably. This drawback can be overcome by modifying the distribution of consecutive slices to the glass slides. Instead of distributing consecutive slices to consecutive slides, consecutive sections will be allocated three times in groups of three slices to three slides in alternating order. This ensures that the expression of each transcript can be judged in nucleus and cytoplasm of one slice whereas the adjacent slices depict the co-localization of the other transcripts in the cytoplasm.

5.4 Hybridization

We usually work with oligonucleotides of about 40–45 nucleotides in length (depending on CG content – see Chapter 1, Section 1.6.3). For the detection of each transcript, the use of at least two different oligonucleotides is recommended (see Chapter 1, Section 1.10). If the respective gene structure is already characterized, one oligonucleotide should be designed to straddle a splicing site (Chapter 1, Fig. 1.6, mRNA2) and the other should be complementary to a sequence within one exon (Fig. 1.6, mRNA1). This ensures that the hybridization is mRNA specific, and identical results from both oligonucleotides serve as control for specificity. Moreover, this design allows for the detection of differentially expressed mRNA splice variants (Laurie and Seeburg, 1994; Laurie *et al.*, 1995; Berthele *et al.*, 1998). The ISH method with oligonucleotide probes as described in Chapter 1 is very specific in the detection of the complementary transcript sequence. From our experience cross-hybridization of a 40mer oligonucleotide probe to similar sequences will usually not occur unless the difference in sequence is less than three nucleotides (see Fig. 5.3B for an example).

ISH of spinal cord sections is carried out routinely in our lab using the protocol outlined in Chapter 1. For spinal cord sections, however, the "minimalist" buffer (Chapter 1, Protocol 1.8) increases background and non-specific binding and should not be used. Employ the "maximalist" buffer (Protocol 1.7).

Slides are removed from ethanol and air-dried; no specific further tissue pretreatment is necessary. 100μl of buffered radiolabelled oligonucleotide is applied to the slides (standard size), which are then hybridized overnight (17 h) at 42°C. After stringent washing, slides are either exposed to X-ray film (e.g. Kodak Biomax) or dipped in photographic emulsion (Kodak NTB2 and Ilford K5 work equally well) as described in Chapter 1 (Protocol 1.9 and Section 1.9.4).

From our experience the ISH method usually works very reliably in the spinal cord. However, in cases where troubleshooting is necessary, we would recommend the following:

- Perform the ISH on corresponding brain slices. This ensures that the oligonucleotide is working.
- Hybridize the spinal cord tissue with other oligonucleotides against housekeeping genes (e.g. GAPDH – see Fig 10.8, chapter 10) to check spinal cord tissue quality.

If mRNA in general is well preserved:

- Increase the concentration of DTT in the probe/buffer mixture (up to 100 mM). This may improve signal-to-noise ratio.
- Incubate for more than 17 h (e.g. 24 h or 48 h). Doing this it is crucial to co-hybridize sections with unlabelled oligonucleotide added in excess, since increases in the duration of incubation may improve the signal but may also increase the non-specific binding of the probe.
- From our experience, changes in the probe concentration do not really help. Changes in hybridization and washing temperatures should be generally avoided, because this only reduces the specificity. Tissue acetylation (Protocol 1.10, Chapter 1) or pretreatment with proteinase do not improve ISH signals in spinal cord tissue from rat or mice, and will even make tissue preservation worse.

After development of emulsion-dipped slides, slices are counterstained with thionin (Protocol 1.9) or cresyl violet. Both dyes work in our hands, but care has to be taken using other dyes since some tend to dissolve the silver grains in a quite unpredictable manner.

5.5 Quantification

The ISH signal in the spinal cord can be quantified macroscopically from X-ray autoradiographs and in cellular resolution from emulsion-dipped sections (see Chapter 9 for a detailed discussion on quantification of ISH data). For semiquantitative measurements of mRNA expression in the spinal cord we analyse at least three sections from each animal. Optical densitometry of ISH X-ray autoradiographs and quantification of silver grain density in cellular resolution is performed using a microscope fitted with a CCD camera and a computer-assisted image analysis system. On X-ray autoradiographs, five areas from each section are chosen for the analysis of the regional distribution of the mRNA transcript investigated: area 1 corresponds to laminae I, II and III, area 2 to laminae II, III and IV (medial edge of the dorsal horn), area 3 to laminae IV and V (lateral edge of the dorsal horn), area 4 to lamina VII of the ventral horn and area 5 to motoneurons in lamina IX (see Fig. 5.2B). Optical densities from each area are measured and averaged using standard procedures.

Quantification of silver grain density in emulsion-dipped sections is of course a much more accurate – nevertheless semiquantitative – measure of the mRNA amount labelled, since this quantification is selective for the ISH label over defined cells or structures (see Chapter 9). To do so within regions of interest (as described above), cell boundaries are delineated by the computer (depending on preselected cell size), or for single cells (e.g. motoneurons) manually on bright-field images. After that, the cell area covered with grains within this boundary is computed on digitized dark-field images and given as a ratio of surface area of the cell covered with grains to the total cell area. Using this method, motoneurons, which have a clear distinction between cytoplasmic and nuclear areas, can even be further analysed concerning the subcellular mRNA expression/segregation to different cellular compartments.

References

Berthele, A., Laurie, D. J., Platzer, S., Zieglgänsberger, W., Tölle, T. R., and Sommer, B. (1998). *Neuroscience* **85,** 733—749.

Berthele, A., Boxall, S. J., Urban, A., Anneser, J. M. H., Zieglgänsberger, W., Urban, L., and Tölle, T. R. (1999). *Dev. Brain Res.* **112,** 39—53.

Brichta, A. M. and Grant, G. (1985). *In* "The Rat Nervous System" (G. Paxinos, Ed), Vol. 2, pp. 293–301. Academic Press, Sydney.

Laurie, D. J. and Seeburg, P. H. (1994). *J. Neurosci.* **14,** 3180–3194.

Laurie, D. J., Putzke, J., Zieglgänsberger, W., Seeburg, P. H., and Tölle, T. R. (1995). *Brain Res. Mol. Brain Res.* **32,** 94–108.

Molander, C., Xu, Q., and Grant, G. (1984). *J. Comp. Neurol.* **230,** 130–141.

Molander, C., Xu, Q., Rivero-Melian, C., and Grant, G. (1989). *J. Comp. Neurol.* **289,** 375–385.

Nicolopoulos-Stournaras, S. and Iles, J.F. (1983). *J. Comp. Neurol.* **217,** 75–85.

Palmer, J. A., De Felipe, C., O'Brien, J. A., and Hunt, S. P. (1999). *Eur. J. Neurosci.* **11,** 3531—3538.

Sommer, B., Keinänen, K., Verdoorn, T. A., Wisden, W., Burnashev, N., Herb, A., Köhler, M., Takagi, T., Sakmann, B., and Seeburg, P. H. (1990). *Science* **249,** 1580–1585.

Tölle, T. R., Berthele, A., Zieglgänsberger, W., Seeburg, P. H., and Wisden, W. (1993). *J. Neurosci.* **13,** 5009–5028.

Tölle, T. R., Berthele, A., Laurie, D. J., Seeburg, P. H., and Zieglgänsberger, W. (1995a). *Eur. J. Neurosci.* **7,** 1235–1244.

Tölle, T. R., Berthele, A., Zieglgänsberger, W., Seeburg, P. H., and Wisden, W. (1995b). *Eur. J. Neurosci.* **7,** 1414–1419.

Wisden, W., Errington, M. L., Williams, S., Dunnett, S. B., Waters, C., Hitchcock, D., Evan, G., Bliss, T. V., and Hunt, S. P. (1990). *Neuron* **4,** 603–614.

Wisden, W., Gundlach, A. L., Barnard, E. A., Seeburg, P. H., and Hunt, S. P. (1991). *Brain Res. Mol. Brain Res.* **10,** 179–183.

CHAPTER 6

PROCESSING HUMAN BRAIN TISSUE FOR *IN SITU* HYBRIDIZATION WITH RADIOLABELLED OLIGONUCLEOTIDES

L. F. B. Nicholson

Department of Anatomy, School of Medicine, University of Auckland, Private Bag 92019, Auckland, New Zealand

6.1 Introduction

The use of *in situ* hybridization to study gene expression in the human brain is vital for gaining a better understanding of the process of neurodegeneration. Newly identified genes stemming from sequencing the human genome could have a possible role in stroke, Parkinson's disease, Alzheimer's disease, Huntington's disease and epilepsy. Expression patterns of candidate genes need to be mapped, first in the healthy brain and then their expression levels and patterns of expression determined in the disease state in which they are implicated. Here I describe the steps useful for preparing human brain samples for the *in situ* hybridization protocol with radiolabelled oligonucleotides (Chapter 1). Published examples using this method to study gene expression in human and monkey brain are: Sirinathsinghji *et al.* (1995), Nicholson and Faull (1996), Rigby *et al.* (1996), Berthele *et al.* (1998, 1999), Bonnert *et al.* (1999), and Schadrack *et al.* (1999). Typical results for $GABA_A$ receptor subunit mRNAs in the human thalamus, hippocampus and neocortex are shown in Fig. 6.1, and metabotropic glutamate receptors in the human cerebellum in Figs 6.2 and 6.3.

INTERNATIONAL REVIEW OF NEUROBIOLOGY, VOL. 47
ISSN/02 $00.00

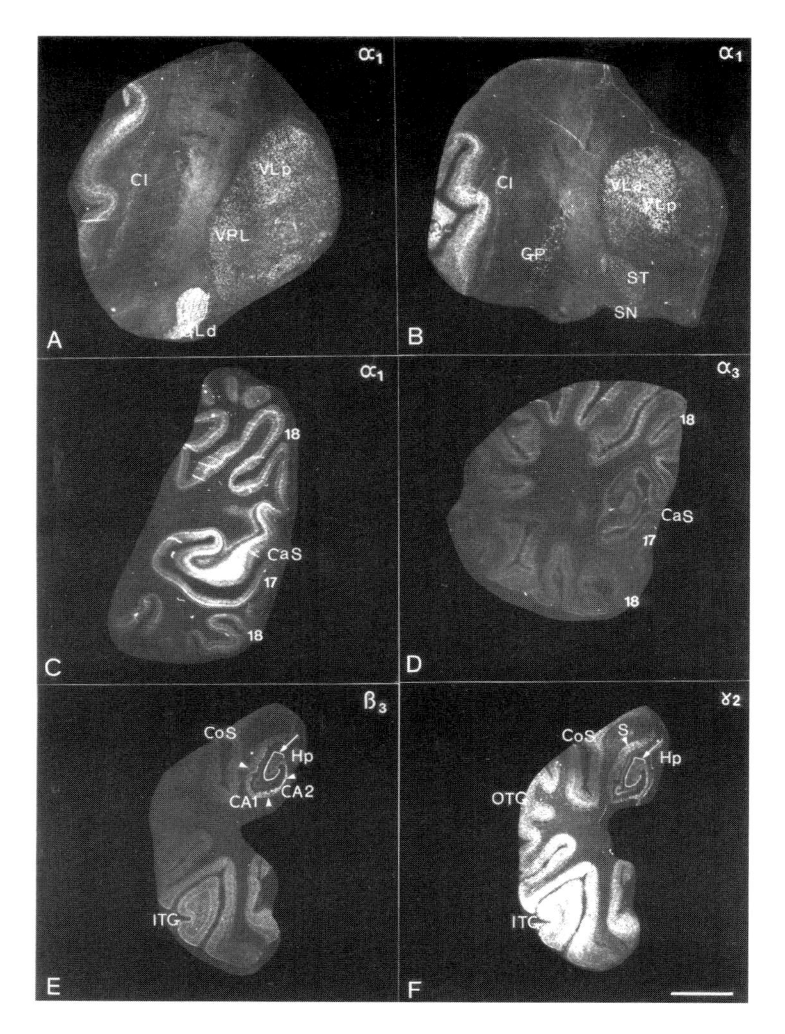

FIG. 6.1. Autoradiographic distribution of $GABA_A$ receptor subunit mRNA in the human brain. Autoradiograms showing the expression of $GABA_A$ receptor mRNA for the α_1 subunit in sections of the human thalamus from different levels of two separate cases (A, H104, level 4; B, H108, level 2), the α_1 (C) and α_3 (D) subunits in sections of the human visual cortex and the β_3 (E) and γ_2 (F) subunits in adjacent sections of human hippocampus all from the same case (HC62). (A) and (B) illustrate the specificity of expression of the α_1 mRNA in the various regions of the thalamus (VLa, VLp, VPL), the subthalamic nucleus (ST), the basal ganglia (GP, SN), and the lateral geniculate (GLd) in two separate cases. (C)–(F) show the expression pattern of various subunits in different regions of the human brain. Whereas α_1 is highly expressed in the primary (C, Brodmann's area 17, CaS) and in the secondary (C, Bodmann's area 18) visual cortex, α_3 is not (D). Similarly, whereas both β_3 and γ_2 are highly expressed in the granular layer of the dentate gyrus of the hippocampus (arrow, E, F), the expression pattern for

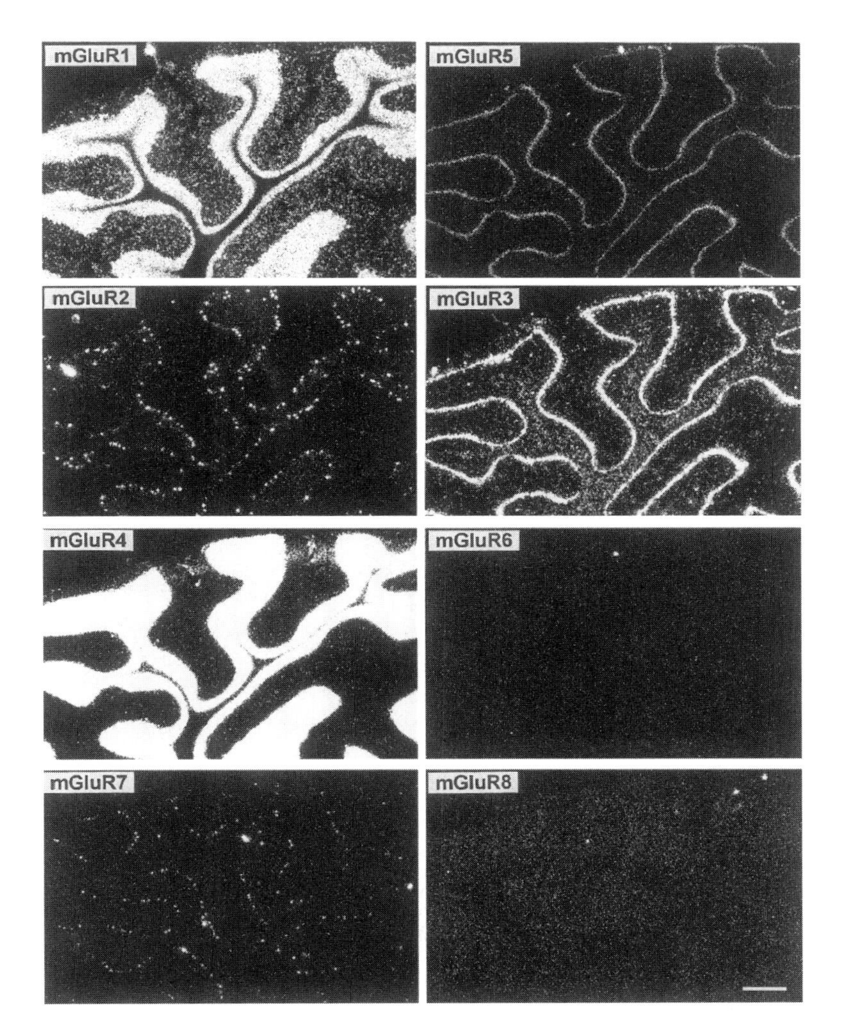

FIG. 6.2. X-ray film autoradiographs showing the mRNA expression of metabotropic glutamate receptor subtypes in the human cerebellar cortex assessed by *in situ* hybridization with subtype-specific ^{35}S-labelled oligonucleotide probes. Scale bar 1 mm. (Reproduced from Berthele *et al.*, 1999.)

FIG. 6.1 *continued*

these subunits in the various regions of Ammon's horn of the hippocampus (Hp) is variable. Subunit β_3 is more highly expressed in the CA1 and CA2 regions, whereas γ_2 is expressed in the subiculum (S). Whereas both subunits show a high level of expression in the inferior temporal cortex (ITG) only γ_2 is expressed in the region of the collateral sulcus (CoS) and the adjacent occipitotemporal gyrus (OTG). Scale bar 1.6 cm. C1, claustrum; GP, globus pallidus, SN, substantia nigra; ST; subthalamic nucleus; CaS, calcarine sulcus.

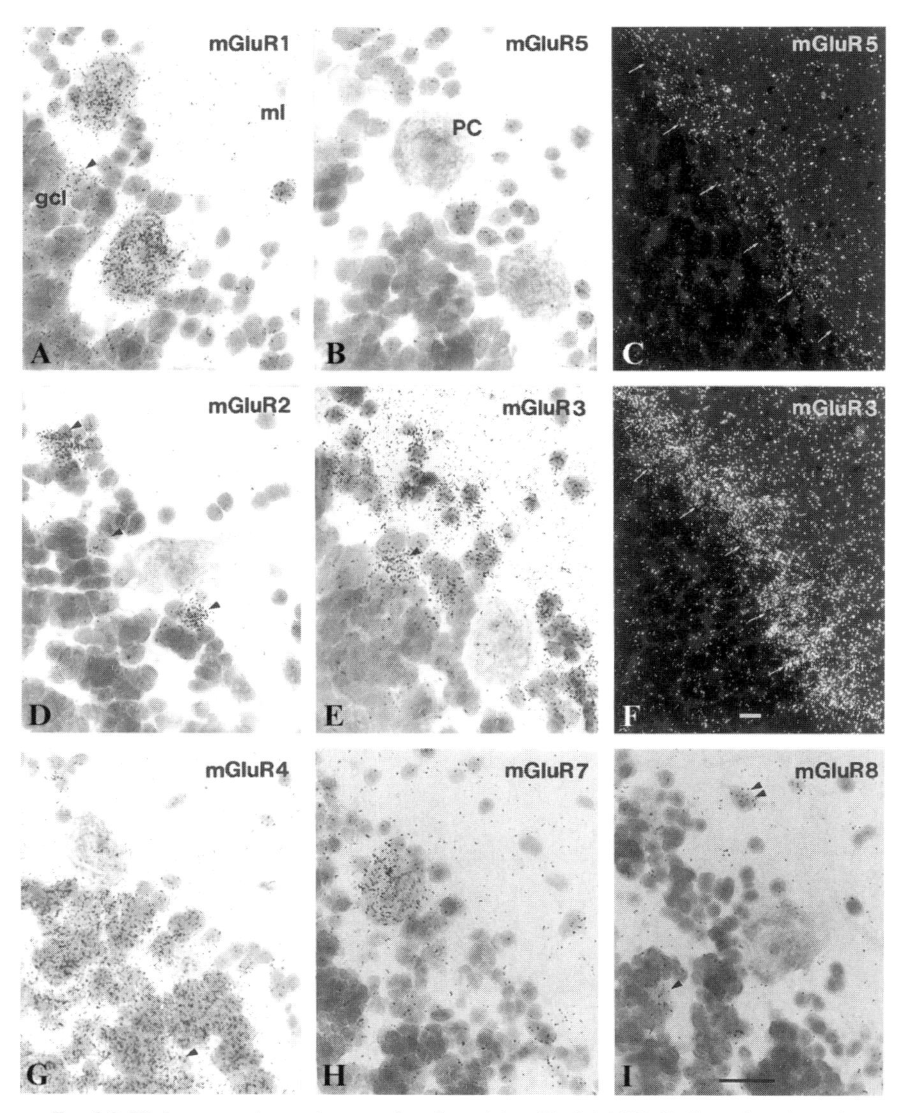

FIG. 6.3. High-power photomicrographs of emulsion (Kodak NTB 2)-dipped sections showing the expression of metabotropic glutamate receptor (mGluR) genes in the human cerebellum at cellular resolution under bright-field (A, B, D, E, G, H, I) and dark-field (C,F) optics; arrowhead, Golgi cell; double arrow head, stellate/basket cell; arrows, Purkinje cell layer. Scale bars 25 μm. Gcl, cerebellar granule cell layer; ml, cerebellar molecular layer; PC, Purkinje cell. (Reproduced from Berthele *et al.*, 1999.)

6.2 Collection of Tissue

6.2.1 Ethical Issues

The use of human tissue for teaching and research purposes falls under strict ethical guidelines, which must be rigorously adhered to. Before embarking on projects involving the use of human tissue you should check carefully the requirements within your own country, as these are sure to vary. In New Zealand we are required by law and Acts of Parliament to obtain the approval of our Institute's Human Ethics Committee, which acts under delegated authority on behalf of the Government. Some countries may, in addition, have State laws or regulations, which must be followed. With the increase in public awareness and the need for researchers to be accountable to the public, it is essential that all ethical issues be addressed before any project begins.

6.2.2 Informed Consent

Human brain tissue is collected from patients who have died from a known neurological disorder and formally bequeathed their brain to research, or from those who have died of accidental death and having no history of any neurological disorder or behaviour. The latter are referred to as "controls". In all cases, consent must be obtained from the next of kin. In the case of a patient with a known neurological disease, we have special donor packages available through the various lay societies, which provide support and advice for patients and their families about the donation of tissue for teaching and research.

6.2.3 Post-mortem Delay and mRNA Stability

Although every effort is made to reduce the post-mortem period, it is now recognized that post-mortem delay is not as critical as was previously thought. Pre-mortem conditions, however, play a significant part in determining the quantity and quality of mRNA (Barton *et al.,* 1993), with post-mortem intervals having a very modest effect on mRNA stability up to at least 36 h after the time of death (Barton *et al.,* 1993). Indeed, a study by Schramm and colleagues found mRNA of three housekeeping genes, as detected by RT-PCR, to be stable for periods of up to 4 days (Schramm *et al.,* 1999). Tissue pH has recently been proposed as a reliable indicator of rapidity of death (RoD) and agonal state, with low pH being associated with prolonged acidosis leading to reduced mRNA, while a high pH is positively correlated with quantifiable amounts of mRNA (Kingsbury *et al.,* 1995; Johnston *et al.,* 1997). Tissue pH has been shown to be stable after death and during freezing/freezer storage (Harrison *et al.,* 1995).

Comparative studies on the effects of post-mortem delay on mRNA stability in rat and human brain tissue found that in the rat, total RNA (quantity) was similar at all post-mortem times (and after freezing) up to 48 h while human brain RNA, although consistent at all post-mortem times, yielded 40–50% that of the rat. Gel blot hybridization showed no degradation of rat brain mRNA, while human cortical RNA appeared slightly degraded. The degree of degradation was not related to post-mortem interval (Johnson *et al.,* 1986).

6.2.4 Removal and Transport of Tissue

Brains must be removed by a skilled mortician to ensure that there is no damage to the cerebrum and that the brainstem remains intact. Blood samples and cerebrospinal fluid required for any diagnostic tests are normally taken at this time. Bequests may come from donors some distance away; brain tissue is best transported on ice (4°C) rather than frozen since it is preferable to block tissue fresh.

6.2.5 Blocking Tissue

We use fresh frozen cryostat sections for our *in situ* hybridization studies but comparable immunohistochemical studies require tissue to be fixed. We therefore normally retain one hemisphere for *in situ* hybridization and ligand-binding studies while the other hemisphere is fixed by perfusion through the cerebral arteries. This tissue can be subsequently blocked and further fixed by immersion.

Fresh brain tissue is blocked in the safety of a biohazards hood. Double gloves and a face shield should be worn since the results of the routine screening tests for hepatitis B and C, HIV and Creutzfeldt-Jakob disease are never back from the laboratory until sometime after the tissue has been received and blocks taken. Cortical and cerebellar blocks are taken at this time for histological processing and diagnostic purposes. Cortical tissue is also taken for the measurement of tissue pH (Johnston *et al.,* 1997) although this can also be done using cerebrospinal fluid (Harrison *et al.,* 1995).

We retain all areas of the brain, taking fresh tissue blocks of approximately 2–4 × 2–3 cm size. It is advisable to follow a simple sequence for blocking, as this will prevent damage to certain regions of the brain that are more difficult to access and block. Cortical blocks are taken first (e.g. three blocks from the motor cortex) and include both the grey and white matter; subsequently the basal ganglia, thalamus and brainstem are taken as coronal slices in a rostral–caudal sequence. These slices include the cingulate gyrus. A useful atlas for identifying and confirming human brain structures is Mai *et al.* (1998). For the selection of tissue blocks, the choice of tissue is, to a large part, dependent on the type of neurological disease under study; for example, the temporal cortex and hippocampus are affected in epilepsy, the striatum in Huntington's disease, while the striatum and substantia nigra are likely to be the areas of interest in Parkinson's disease.

Once blocked, tissue is immediately frozen under powdered dry ice. We place the blocks into small, clean metal dishes and sieve the dry ice over the tissue. After standing for 2–3 min the blocks should be frozen (the exact time depends on the size of the block) and then removed from the tray and wrapped in a double layer of tin foil, clearly and carefully labelled, placed in a sealed plastic bag, and subsequently stored in the –80°C freezer. Although long-term freezing and storage of tissue may affect the integrity of mRNA, studies by Leonard and colleagues found that even after storage for 5 years there was enough full-length and partial transcripts present to be amplified using PCR (Leonard *et al.*, 1993). Other studies, using shorter time intervals, have found no correlation between freezer storage and RNA stability or content (Harrison *et al.*, 1995).

6.3 Cryostat Sectioning

6.3.1 Safety Precautions

When cutting sections of human brain tissue we take the precautionary approach and follow the standard procedures of wearing laboratory coats and double gloves. Tissue blocks should not, however, be sectioned until all laboratory results have been received. Tissue must be discarded if laboratory results are positive.

In multi-user facilities, where the cryostat is not dedicated to sectioning human brain tissue, a logbook should record the type of tissue being cut and the type and level of any infection (if known).

The cryostat and microtome should be disinfected and sterilized (Protocol 6.1):

- As part of a normal maintenance regime
- Before servicing (it is now standard practice for service agents to require a signed certificate stating that the cryostat has been disinfected)
- Immediately after cutting tissue known to be infected.

Traditionally, diagnostic laboratories have sterilized their cryostat and microtome using a potassium permanganate and formaldehyde vapour mix (see Protocol 6.1). In the past cryostats were often cleaned with a 2% glutaraldehyde solution, which was liberally "splashed" over the inside of the cryostat and the microtome and left for 30 min before being drained away and the cryostat wiped out and dried. Although the disinfectant properties of glutaraldehyde are acknowledged, it is a very toxic substance and a strong carcinogen. For obvious health and safety reasons glutaraldehyde should not be used as a disinfectant.

Both of these methods require the cryostat to be turned off (i.e. defrost mode) and this takes a considerable amount of time. In the past, this was the only reliable method of sterilization. There is now, however, an excellent product on the market called Cryofect (Leica Microsystems, http://www.hbu.de/products/cryofect.htm), a surface disinfectant spray

---

PROTOCOL 6.1 DISINFECTION OF CRYOSTAT.

1. Switch off electric supply.
2. Weigh 2.4 g of potassium permanganate into a plastic weighing boat and transfer to a beaker. Take to cryostat and add 4.4 ml of 40% formaldehyde; wait for reaction to start.
3. Once the reaction has started, place immediately in the cryostat, close the lid and leave overnight.
4. Quench the residual formalin vapour by placing 2 ml of concentrated ammonia onto a paper towel inside the cabinet. Leave 1 h.
5. The cabinet can now be cleaned of debris etc. Place microtome knife in box. (Resharpen before replacing.)
6. Remove the microtome and ensure it is dry. Lubricate moving parts with low-temperature oil.
7. Ensure cabinet is dry before replacing microtome. (A hairdryer is helpful at this stage.)
8. Replace the knife and align anti-roll plate.
9. Switch on the electric supply.
10. Check that the automatic defrost clock is set to the correct time.
11. As soon as the working temperature of approximately –20°C has been reached check that microtome is functional.

---

that is a reliable virucidal and germicide agent, is safe to use and is effective against bacteria, fungi and viruses including HBV, HIV, herpes and hepatitis B (hepatitis C trials are still being carried out). There is extensive literature available from Leica on the viral and bacterial efficacy of Cryofect and on the recommended safety precautions for users. Cryofect contains ethanol as the active ingredient (72%) plus 0.16% chlorhexidindigluconate. It is highly flammable so needs to be kept away from ignition sources. Spills can be washed away with water and the usual procedures followed for skin and eye contact (see www.hbu.de/products/cryofect.htm).

Cryofect is available in Europe and Asia Pacific, but due to EPA regulations further tests are required before its release onto the USA market. At present, the traditional method of vapour disinfectant is the only option.

6.3.2 Matching Controls

Since changes in mRNA expression in tissue from patients who have died with a known neurological disorder is the principal focus of our research, we use the patient details as our "fixed" criteria and then try to match this as closely as possible with equivalent control tissue from donors who have died of a non-neurological disease.

Although mRNA is stable in human brain tissue with long periods of post-mortem delay and so stability is not an important consideration with respect to short post-mortem times, we still attempt to age, sex and post-mortem delay match our test and control cases; tissue from younger cases with short post-mortem times is preferable. Indeed Harrison *et al.* (1995), in their study of the effects of pre-mortem acidosis and post-mortem delay on RNA degradation, found that brain tissue pH decreased with increased age at death, was directly related to agonal state at death, but was independent of post-mortem interval and the presence of hypoxic changes. Obviously there are a variety of pre- and post-mortem factors that can affect mRNA levels to varying degrees; some we can attempt to control for (age, sex, post-mortem delay) while others (acidosis and hypoxia) are much more difficult. Nonetheless, we do need to be aware of these variables.

6.3.3 Sectioning

Tissue blocks should be taken from the –80°C freezer and allowed to equilibrate to cryostat chamber temperature (–20°C) for 12 h; overnight is normally an adequate and convenient time interval. Blocks are mounted onto large cryostat chucks using Tissue-Tek (Miles Scientific) and sections cut at 16 μm. Slide size is dependent on the size of the block. Routinely we use 76 mm × 26 mm (Lomb Scientific), 76 mm × 38 mm (Sail Brand) and for extra-large sections 76 mm × 48 mm and 76 mm × 58 mm (Blue Star). All slides are 1–1.2 mm thick. Sections are then thaw-mounted onto cleaned, double-dipped, chrome alum gelatine-coated slides. We find the chrome alum gelatine coating reliable but poly-L-lysine coating (Chapter 1, Protocol 1.1) is an alternative. Sections are normally quite large and require skilled handling to ensure they are mounted flat; folds tend to allow solutions to get underneath the tissue and this in turn can cause the sections to lift off the slide during high-stringency post-hybridization washes. A description of the general features of cryostats and more on cyrostat sectioning, including troubleshooting, can be found in Chapter 2.

6.4 Procedures for Using Radioactive Oligonucleotide Probes

6.4.1 Pre-treatment of Sections

The protocol for radioactive oligonucleotide *in situ* hybridization outlined in Chapter 1 of this book has proven to be excellent for studies on human tissue. We suggest just a few minor modifications. We always pre-treat our human brain sections with 0.25% acetic anhydride and dehydrate in 100% chloroform (as outlined in Chapter 1, Protocol 1.10) since human brain tissue has a high lipid content and can give high non-specific background especially in the white matter. These steps certainly help to reduce the non-specific binding. Similar considerations apply when hybridizing ^{35}S-labelled probes to monkey brain sections (Sirinathsinghji, 1994).

6.4.2 Hybridizing and Washing

There are some minor variations in the composition and final concentrations of the components of our hybridization buffer when compared with that given in Protocols 1.7 and 1.8 of Chapter 1: we use 10-fold less Denhardt's solution (0.5 × Denhardt's solution); and twice the amount of dextran sulfate (20% dextran sulfate) in the hybridization buffer (see Nicholson and Faull, 1996). We routinely apply 100 μl of probe/hybridization buffer to large sections. However, for sections that are mounted on extra-large slides (58 mm × 76 mm) we apply 150 μl of the probe/hybridization buffer. One probe labelling/experiment is normally sufficient.

When using oligonucleotide probes with 100% base pair identity to human mRNA we hybridize at 43°C but carry out the post-hybridization washes at 45°C. With probes that may have originally been designed for use on rat or mouse tissue, but which have high sequence identity with the known human sequence (say 1–4 base pair "wobble"), we decrease the hybridizing temperature but still carry out the post-hybridization washes at 45°C. While this washing temperature is lower than the 55°C stringency wash in 1 × SSC recommended in Protocol 1.6, we follow 2 × 15-min washes in 1 × SSC with 2 × 15-min washes in 0.5 × SSC (higher stringency) then 1 × 15-min 0.5 × SSC wash at room temperature. DTT (10 mM) is added to all washes, as experience has shown that this does reduce the background in human tissue sections. Ammonium acetate (300 mM) is added to the 70% ethanol as part of the final dehydration as this tends to enhance the signal.

Sections can then be exposed to X-ray film (e.g. Kodak Biomax, Chapter 1, Section 1.9 and Figs 6.1 and 6.2) or emulsion dipped (Fig. 6.3) (Chapter 1, Protocol 1.9). For emulsion dipping large slides, we use a perspex dipping chamber that accommodates the extra width of the slide; this was custom made in our faculty workshop.

Acknowledgements

Human tissue is bequeathed to the New Zealand Neurological Foundation Human Brain Bank. We are grateful to the New Zealand Neurological Foundation for their support and to Professor R. L. M. Faull for the establishment and directorship of this invaluable facility. Christine Tait must be commended for her tenacious perfection of the *in situ* methods on human sections.

References

Barton, A. J., Pearson, R. C., Najlerahim, A., and Harrison, P. J. (1993). *J. Neurochem.* **61,** 1–11.

Berthele, A., Laurie, D. J., Platzer, S., Zieglgänsberger, W., Tölle, T. R., and Sommer, B. (1998). *Neuroscience* **85,** 733-749.

Berthele, A., Platzer, S., Laurie, D. J., Weis, S., Sommer, B., Zieglgänsberger, W., Conrad, B., and Tölle, T. R. (1999). *NeuroReport* **10**, 3861–3867.

Bonnert, T. P., McKernan, R. M., Farrar, S., le Bourdelles, B., Heavens, R. P., Smith, D. W., Hewson, L., Rigby, M. R., Sirinathsinghji, D. J., Brown, N., Wafford, K. A., and Whiting, P. J. (1999). *Proc. Natl Acad. Sci. USA* **96**, 9891–9896.

Harrison, P. J., Heath, P. R., Eastwood, S. L., Burnet, P. W., McDonald, B., and Pearson, R.C. (1995). *Neurosci. Lett.* **200,** 151–154.

Johnson, S. A., Morgan, D. G., and Finch, C.E. (1986). *J. Neurosci. Res.* **16,** 267–280.
Johnston, N. L., Cerevnak, J., Shore, A. D., Torrey, E. F., Yolken, R. H., and The Stanley Neuropathology Consortium (1997). *J. Neurosci. Methods* **77,** 83–92.
Kingsbury, A. E., Foster, O. J. F., Nisbet, A. P., Cairns, N., Bray, L., Eve, D. J., Lees, A. J., and Marsden, C. D. (1995). *Mol. Brain Res.* **28,** 311–318.
Leonard, S., Logel, J., Luthman, D., Casanova, M., Kirch, D., and Freedman, R. (1993). *Biol. Psychiatry* **33,** 456–466.
Mai, J. K., Assheuer, J., and Paxinos, G. (1998). "Atlas of the Human Brain". Academic Press, London.
Nicholson, L. F. B. and Faull, R. L. M. (1996). *In* "The Basal Ganglia V" (Ohye *et al.*, Eds), pp. 433–439. Plenum Press, New York.
Rigby, M., Le Bourdelles, B., Heavens, R. P., Kelly, S., Smith, D., Butler, A., Hammans, R., Hills, R., Xuereb, J. H., Hill, R. G., Whiting, P. J., and Sirinathsinghji, D. J. (1996). *Neuroscience* **73,** 429–447.
Schadrack, J., Willoch, F., Platzer, S., Bartenstein, P., Mahal, B., Dworzak, D., Wester, H. J., Zieglgänsberger, W., and Tölle, T. R. (1999). *NeuroReport* **10,** 619–624.
Schramm, M., Falkai, P., Tepest, R., Schneider-Axmann, T., Przkora, R., Waha, A., Pietsch, T., Bonte, W., and Bayer, T. A. (1999). *J. Neural Transm.* **106,** 329–335.
Sirinathsinghji, D. J. S. (1994). *In* "*In Situ* Hybridization Protocols for the Brain", first edition (W. Wisden and B. J. Morris, Eds), pp. 41–43. Academic Press, London.
Sirinathsinghji, D. J., Rigby, M., Heavens, R. P., Smith, D., Fernandez, J. M., Schuligoi, R., and Hill, R. G. (1995). *Neuroscience* **65,** 51–57.

CHAPTER 7

IN SITU HYBRIDIZATION OF ASTROCYTES AND NEURONS CULTURED *IN VITRO*

L.. Ariza-McNaughton,* C. De Filipe** and S. P. Hunt†

*Vertebrate Development Laboratory, Cancer Research UK, PO Box 123, 44 Lincoln's Inn Fields, London WC2A 3PX, UK

**Instituto de Neurociencias, University Miguel Hernandez, San Juan, E-03550, Alicante, Spain

†Department of Anatomy and Developmental Biology, University College London, Gower Street, London WC1E 6BT, UK

7. 1 Introduction

Many advantages are offered by the use of isolated cells in culture. It is possible to obtain pure cell populations and control their growth, differentiation or phenotype depending on the culture conditions. Furthermore, cells in culture constitute a powerful tool for the localization and quantification of gene expression in mixed populations of cells. The mechanisms of regulation of gene expression following stimulation can easily be studied in this controlled system (McNaughton and Hunt, 1992; De Felipe and Hunt, 1994; Morris, 1995, 1997; Simpson and Morris, 1995a, b, 2000; Morris *et al.*, 2000; Kenrick *et al.*, 2001). When mixed cultures are required for studies of interactions between different cell types (e.g. astrocytes or Schwann cells with central or peripheral neurons), *in situ* hybridization (ISH) allows the study of the differential gene expression pattern of the two populations in response to a given stimulus (De Felipe *et al.*, 1993; Simpson and Morris, 1995b). In addition, when cells are cultured on specialized microplates, *in situ* hybridization can be linked to high-throughput screening systems (Kenrick *et al.*, 2001).

In this chapter we describe techniques for the localization of mRNAs in primary cultures of neurons and glial cells, using ^{35}S-labelled oligonucleotide probes. We limit our chapter to the special issue of methodology for

INTERNATIONAL REVIEW OF NEUROBIOLOGY, VOL. 47
ISSN/02 $00.00

ISH to astrocytes, Schwann cells or neurons in culture. However, the method can also be applied to other cell types grown in culture. We do not discuss the probe labelling, hybridization buffer or washing conditions as these are referred to in detail in Chapter 1.

We use two methods: (i) the growth of cells on glass coverslips and (ii) the growth of cells in chambers mounted on to glass slides (Lab-Tek/Gibco chamber/slide, supplied by Lab-Tek Division, Miles Laboratories, Inc., 30W475 North Aurora Road, Naperville, IL 60540, USA). There are advantages and disadvantages in the use of the Lab-Tek chamber/slides or the traditional coverslips. The tissue culture chambers are costly in comparison with ordinary coverslips, although they are very much easier to handle, because, after the culturing period, they can be treated and manipulated like slides with brain sections on them (see Chapter 1). The Lab-Tek chambers also enable one differentially to stimulate different groups of cells on the same slide, add different substances to one chamber of each cell type (see McNaughton and Hunt, 1992), and also enable the simultaneous hybridization of different probes to the same slide. However, in our hands, some neuronal cell types such as dorsal root ganglion cells are difficult to grow in the Lab-Tek chambers. In this case, we recommend the use of traditional glass coverslips.

7.2 Hybridization Procedures for Lab-Tek Tissue Culture Chamber/Slides

Astrocytes are isolated as described by Lim and Bosch (1990) and plated on to poly-L-lysine- (Sigma; catalogue no. P9155) or poly-DL-ornithine- (Sigma; catalogue no. P8638) coated glass tissue culture chamber/slides (Lab-Tek/Gibco). Glass slides must be used, as plastic ones will dissolve in the Histoclear or xylene used for coverslipping them. Further, plastic slides sometimes float out of the carrying racks to the surface of the processing solutions, making them very difficult to handle and store. Before the cells are plated, the wells should be coated with 100 μl of poly-DL-ornithine or poly-L-lysine and left for 2 h at room temperature. They are then rinsed twice with 1 ×γ PBS and the cells are plated. (Poly-DL-ornithine at 0.01% is dissolved in 0.15 M sodium borate buffer pH 8.4. This solution can be kept at 4°C for several months. Poly-L-lysine can be made up in water as stock and kept frozen at –20°C for months.)

For our particular experiments (see McNaughton and Hunt, 1992), the cells are usually grown for 8–10 days before they are challenged with the different stimuli. The Lab-Tek culture chamber/slides come with two, four or

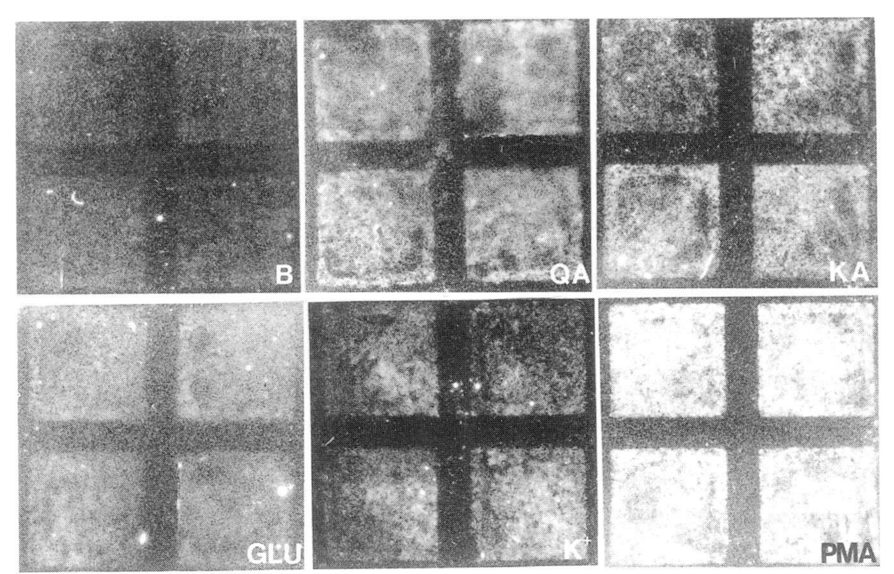

FIG. 7.1. Dark-field photomicrograph of cultured astrocytes hybridized with a c-*fos*-specific oligonucleotide. Treatments were as follows: B, unstimulated; QA, quisqualic acid, 100 μM; KA, kainic acid, 100 μM; GLU, glutamic acid, 100 μM; K^+, 140 μM; PMA, phorbol ester, 200 nM (see McNaughton and Hunt, 1992). (Reproduced with permission from Elsevier Science Publishers.)

eight chambers/wells mounted on to each glass slide. Thus, different treatments or stimuli can be applied to the same slide (Fig. 7.1) and consequently processed in parallel, so that intra-assay variations are minimized. For most of our studies, 20–30 min of stimulation with excitatory amino acid was long enough to see changes in immediate early gene expression for cells in culture (see Fig. 7.1 and McNaughton and Hunt, 1992). However, 12–24 h or longer stimulations have also been carried out (De Felipe *et al.*, 1993; Morris, 1997).

After the culture period, the plastic walls and seal that compartmentalize the slide can be easily removed according to the manufacturer's instructions and the slide is ready to be processed in exactly the same way as slides with brain sections using the staining troughs and racks illustrated in Chapter 1 (Fig. 1.4); i.e. the slides are rinsed in 250 ml of DEPC-treated 1 × PBS (see Chapter 1, Protocol 1.2, footnote *c*, for PBS recipe). Cells are then fixed for 5 min in 250 ml of cold 4% paraformaldehyde in 1 × PBS. After the fixation the cells are rinsed in 250 ml of PBS followed by dehydration through graded alcohols (Chapter 1, Protocol 1.2). This procedure is carried out at room temperature. Samples are ready to be hybridized or alternatively can be stored long-term in 95% ethanol at 4°C until required for hybridization (see Chapter 1). For the hybridization, 100 µl of hybridization buffer per probe per slide is added with a parafilm coverslip (Chapter 1, Protocol 1.5).

Alternatively, different probes can be used in the same slide if the upper structure which divides the slide is not removed. In this case, 100 µl of the probe hybridization buffer is added to each well. No parafilm is necessary. Hybridization is performed for 20–24 h at 42°C (Chapter 1, Protocol 1.5, for probe concentration and hybridization buffer). After hybridization, the chambers can be removed from the slides and they can be washed as normal (Chapter 1, Protocol 1.6).

7.3 Hybridization Procedures for Glass Coverslips

Adult sensory neurons are isolated as described by Lindsay (1988) and plated on to poly-L-lysine- or poly-DL-ornithine-coated 19 mm round glass coverslips (BDH/Merck no. 1), which are placed in 12-well plates (made by Linbro or Costar). As for the Lab-Tek slides, glass coverslips must be used as plastic ones will dissolve in the Histoclear or xylene used for coverslipping them. Laminin (Sigma; catalogue no. L2020), an alternative substrate used in some neuronal cultures, does not interfere with the ISH technique and therefore can be used if needed.

Throughout the procedure, coverslips are manipulated by the combined action of a pair of fine forceps and a fine needle (e.g. syringe or mounting pin), the needle being used to help tease and lift the coverslips from the bottom of the well. They are generally not removed from the wells, and solutions (usually between 3 and 4 ml per well) are pipetted on to and aspirated off them. In the cases where the coverslips are removed, for example in order to dry them after ethanol storage or after the post-hybridization washing steps, they can be placed in the type of rack illustrated in Fig. 7.2.

Before plating and culturing of the cells, the coverslips are washed in HCl (0.5 M) for 1 h, after which they are rinsed in copious distilled water with a final rinse in absolute (100%) ethanol. The coverslips are allowed to dry, wrapped in aluminium foil and sterilized by baking for 2 h at 180°C. The washed coverslips are placed in 12-well plates using forceps. Then 1 ml per well of 0.01% poly-DL-ornithine is added, and the coverslips are left for 2 h at room temperature followed by a rinse with 1 × PBS. Then 0.5 ml of laminin (5 µg/ml in 1 × PBS stocked in aliquots at –70°C) is immediately

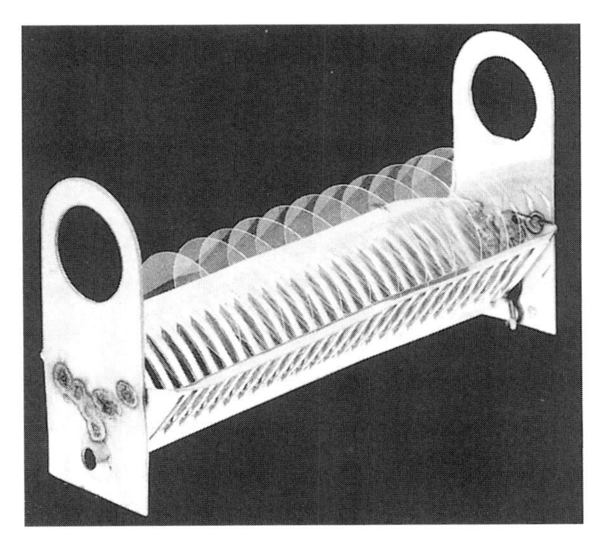

FIG. 7.2. Illustration of the type of rack used for holding the glass cell-culture coverslips.

added, and the mixture incubated at 37°C for 3 or 4 h. The laminin is aspirated (the coverslips must not be rinsed) and the cells are plated on.

After the culture period, the coverslips are rinsed in 2 ml of DEPC-treated 1 × PBS. Cells are then fixed for 5 min in 2 ml of cold 4% paraformaldehyde in 1 × PBS (Chapter 1). After fixation, the cells are rinsed in 1 × PBS followed by dehydration through graded alcohol. At each stage, 2 ml of each solution is pipetted on to and then aspirated from the coverslips. At no point are the coverslips removed from the wells. This procedure is carried out at room temperature. Samples are ready to be hybridized or stored in 95% or absolute ethanol at 4°C (Chapter 1, Protocol 1.2). For long-term storage, 3 or 4 ml of ethanol is pipetted into each well. The lid is sealed with parafilm, and the plates placed at 4°C.

Before hybridization, coverslips are removed from the dish and air-dried in a rack (Fig. 7.2) in a dust-free environment at room temperature for 30 min. The coverslips are then placed back in a new clean sterilized 12-well plate, where the next steps of hybridization and washing are actually carried out.

The probe to be hybridized is added in 50 µl of hybridization buffer (probe concentration and hybridization buffer composition as in Chapter 1) and pipetted on to the coverslips. This is then covered with hand-cut parafilm coverslips cut into small circles, avoiding the formation of air bubbles between the coverslip and the slide. The dishes are incubated at 42°C

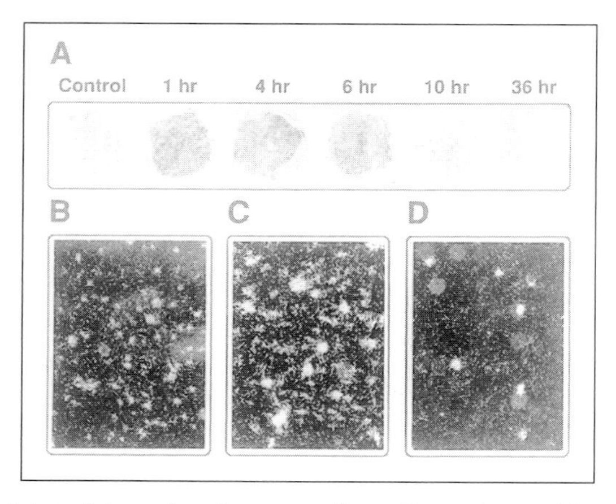

FIG. 7.3. Dissociated adult rat dorsal root ganglion cells growing on 19 mm coverslips. Cells were treated with cycloheximide, fixed at different times and hybridized with ^{35}S-labelled *c-jun* antisense oligonucleotide. (A) Autoradiograph image of the coverslips. (B), (C) and (D) silver grains over neurons (large cells) and Schwann cells (small cells) under dark-field illumination in control cells (B) or after 2 h (C) or 12 h (D) of treatment respectively. *c-jun* mRNA was increased from 1 to 8 h of treatment in both cell types. After 10 h of treatment and onwards *c-jun* expression disappeared in Schwann cells but not in neurons (see De Felipe and Hunt, 1994).

in a humidified atmosphere for 18–20 h. It is important to leave at least two empty wells in the dish and to add distilled water to them. This keeps a humidified atmosphere in the dish. Any dehydration of the sample will cause erratic and false results because the hybridization and washes are carried out in an air incubator.

After hybridization, covering the coverslips with liquid makes them much easier to remove, therefore 4 ml of pre-warmed 1 × SSC (55°C) is pipetted into each well before removal of the parafilm coverslips. Then, by combined teasing with forceps and a syringe needle, the parafilm coverslips can be easily removed. The wells are placed in an air incubator set at the washing temperature (55°C). After half an hour, the SSC is changed by aspiration, and another 4 ml of pre-heated 1 × SSC is pipetted into each well. After the final washing step, the coverslips are removed one at a time from the wells with forceps and needle and dipped sequentially into 0.1 × SSC, 70% ethanol and 95% ethanol. The coverslips are air-dried in racks (Fig. 7.2).

With the cells facing up, the coverslips are then attached to a cleaned histology slide by using DPX (BDH) as glue, so that the slides can be either exposed to X-ray film (Kodak, Fig. 7.3) and/or dipped in emulsion (Figs 7.1, 7.3 and 7.4). X-ray film is very useful for macroscopic observations and

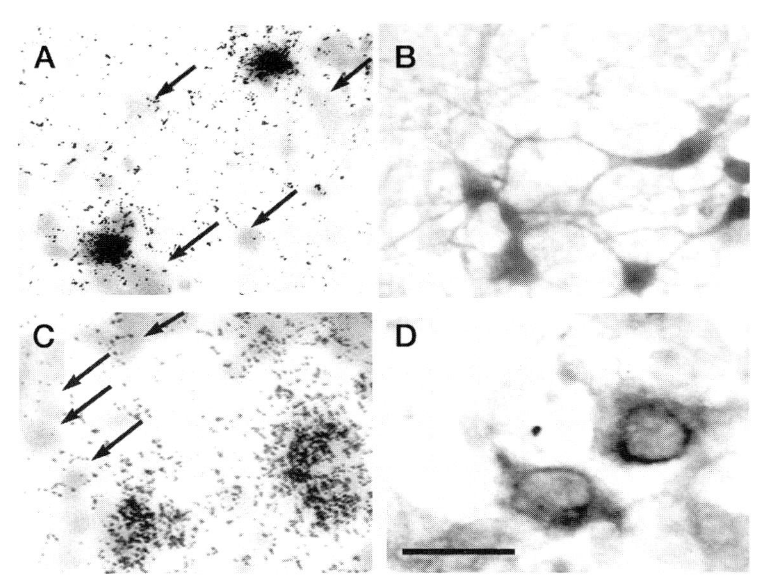

FIG. 7.4. Cellular expression of mRNAs in cultured neurons and glia. (A) ISH signal from a mRNA (in this case somatostatin mRNA) expressed in a subpopulation of cultured rat striatal neurons – two neurons are labelled with silver grains, while other counterstained neurons (arrows) are not labelled. Exposure time three weeks. (B) Parallel culture processed for immunocytochemistry to detect neuron-specific enolase, illustrating some of the distinctive morphologies of neurons in culture. (C) ISH signal from a mRNA (in this case zif/268 mRNA) expressed in glia – two astrocytes are labelled with silver grains, while counterstained neurons (arrows) are not labelled. Note that astrocytic labelling is readily identifiable due to the distinctive large, flat morphology of astrocytes in culture. Exposure time five weeks (see Simpson and Morris, 1995a and b for further details). (D) Parallel culture processed for immunocytochemistry to detect glial fibrillary acidic protein (GFAP). The ISH was performed using ^{35}S-labelled oligonucleotides, and autoradiographic detection was accomplished using Ilford K5 emulsion. Scale bar 40 μm.

quantification of hybridization with radiolabelled oligonucleotide probes, where different levels of expression can be assessed by densitometry (see Chapter 9). When a more detailed and fine analysis is needed, dipping for autoradiography in photographic emulsions is the method of choice (Chapter 1, Protocol 1.9). The final thionin staining allows identification of the cellular localization of the silver grains.

References

De Filipe, C. and Hunt, S. P. (1994). *J. Neurosci.* **14,** 2911–2923.

De Filipe, C., Jenkins, R., O'Shea, R., Williams, T. S. C., and Hunt, S. P. (1993). *Adv. Neurol.* **59**, 263–271.

Kenrick, M., Potts, C., and Turner, J. (2001). Amersham Pharmacia Life Science News, http://www.apbiotech.com/product/publication/Isn/Isn-1-8/lsn-1-8.htm

Lim, R. and Bosch, E. P. (1990). *In* "Methods in Neurosciences, Vol. 2, Cell culture" (P.M. Conn, Ed.). Academic Press, New York.

Lindsay, R. M. (1988). *J. Neurosci.* **8,** 2394–2405.

McNaughton, L. A. and Hunt, S. P. (1992). *Mol. Brain Res.* **16,** 261–266.

Morris, B. J. (1995). *J. Biol. Chem.* **270,** 24740–24744.

Morris, B. J. (1997). *Eur. J. Neurosci.* **9,** 2334–2339.

Morris, B. J,. Newman-Tancredi, A., Audinot, V., Simpson, C. S., and Millan, M. J. (2000). *J. Neurosci. Res.* **59,** 740–749.

Simpson, C. S. and Morris, B. J. (1995a). *Neuroscience* **68,** 97–106.

Simpson, C. S. and Morris, B. J. (1995b). *Neuropharmacology* **34,** 515–520.

Simpson, C. S. and Morris, B. J. (2000). *J. Biol. Chem.* **275,** 16879–16884.

CHAPTER 8

IN SITU HYBRIDIZATION ON ORGANOTYPIC SLICE CULTURES

A. Gerfin-Moser* and H. Monyer**

*Schweizerische Multiple Skelerose Gesellschaft, Brinerstrasse 1, Postfach, 8036 Zürich, Switzerland

**Department of Clinical Neurobiology, University of Heidelberg Im Neuenheimer Feld 364, D-69120 Heidelberg, Germany

8.1 Introduction

Slice cultures of nervous tissue represent an *in vitro* preparation that is thin enough to use for experimental manipulations and visualization of individual cells, yet retains much of the basic structural and connective organization of its tissue of origin (Gähwiler *et al.*, 1997). In culture, the slices become thinned to a quasi-monolayer, facilitating study of the single cells that form the intricate neuronal network. Organotypic neuronal cultures can be derived from various brain regions. The cytoarchitectural organization of the parent slice is retained and the spatial distribution of synaptic connections, which develop *"de novo"* during the first weeks in culture, forms much as it would *in situ*. As an *in vitro* analogue of axonal connectivities between brain areas, slices from regions that are anatomically remote but interconnected in the brain can be co-cultured, providing a powerful model for studying the development and function of neuronal projections (Gähwiler *et al.*, 1991, 1997). Organotypic cultures thus offer several advantages over other preparations, such as cultures of dissociated cells

INTERNATIONAL REVIEW OF NEUROBIOLOGY, VOL. 47
ISSN/02 $00.00

or acute slices, and are well suited for acute and long-term pharmacological, electrophysiological and morphological studies. For example, it is of interest to study changes in gene expression associated with chronic epilepsy (Müller *et al.*, 1993; Moser *et al.*, 1993), growth factors (Gähwiler *et al.*, 1987) or developmental changes in the culture system (Finsen *et al.*, 1992).

In situ hybridization (ISH) allows the visualization of gene expression in cells at the mRNA level. Each cell represents a living single unit within a highly organized network, which can be investigated under standard culture conditions and also after a variety of experimental manipulations. In this chapter we illustrate the use of the ISH technique, based on the procedure described in Chapter 1 and adapted for organotypic neuronal cultures (Gerfin-Moser *et al.*, 1995). We have used probes for several ligand-gated ion-channel receptor subunits on rat hippocampal slice cultures which were allowed to mature under standard conditions for 2–3 weeks (Gerfin-Moser *et al.*, 1995). The cultures were kindly provided by the laboratory of Dr B. H. Gähwiler at the Brain Research Institute in Zurich. We chose oligonucleotide probes not only because they allow for subunit specificity but also because they yield lower non-specific labelling than cRNA probes. This latter point was particularly important with respect to the labelling obtained in the pyramidal cell regions of the hippocampal cultures as they maintained a thickness of two to three cell layers. The hybridization pattern was revealed by exposure to X-ray film or autoradiographic emulsion as stated. Another excellent example of the use of ^{35}S-labelled oligonucleotides hybridized to hippocampal slice cultures comes from Bauer *et al.* (2000), who looked at factors regulating chromogranin mRNA levels.

As shown in Fig. 8.1 (A,D,G), Nissl staining revealed a certain variability in the shape of the cultured hippocampal formation due to various amounts of cell migration and flattening of the culture compared with the size and shape of the original slice. Hybridization with probes for the AMPA-selective glutamate receptor subunits A or B of the rat (GluR-A, GluR-B) (Boulter *et al.*, 1990; Keinänen *et al.*, 1990) gave a strong signal in all hippocampal areas for both probes, as shown by the corresponding images of the X-ray film (Fig. 8.1B,E). On the other hand, labelling of the cells with a probe specific for the subunit γ_2 of the $GABA_A$ receptor resulted in a much fainter image of the hippocampal formation, although the specific radioactivity of the probe used was similar to those for GluR-A and -B and the exposure to X-ray film was 2 days longer (7 days; Fig. 8.1H). In this case, the relative signal intensities corresponded roughly to what would be expected from comparison with data obtained in brain sections (Killisch *et al.*, 1991; Werner *et al.*, 1991). Control hybridization of cultures with a mixture containing the radioactively labelled probe and an excess of the unlabelled oligonucleotide suppressed signals almost completely for all three probes

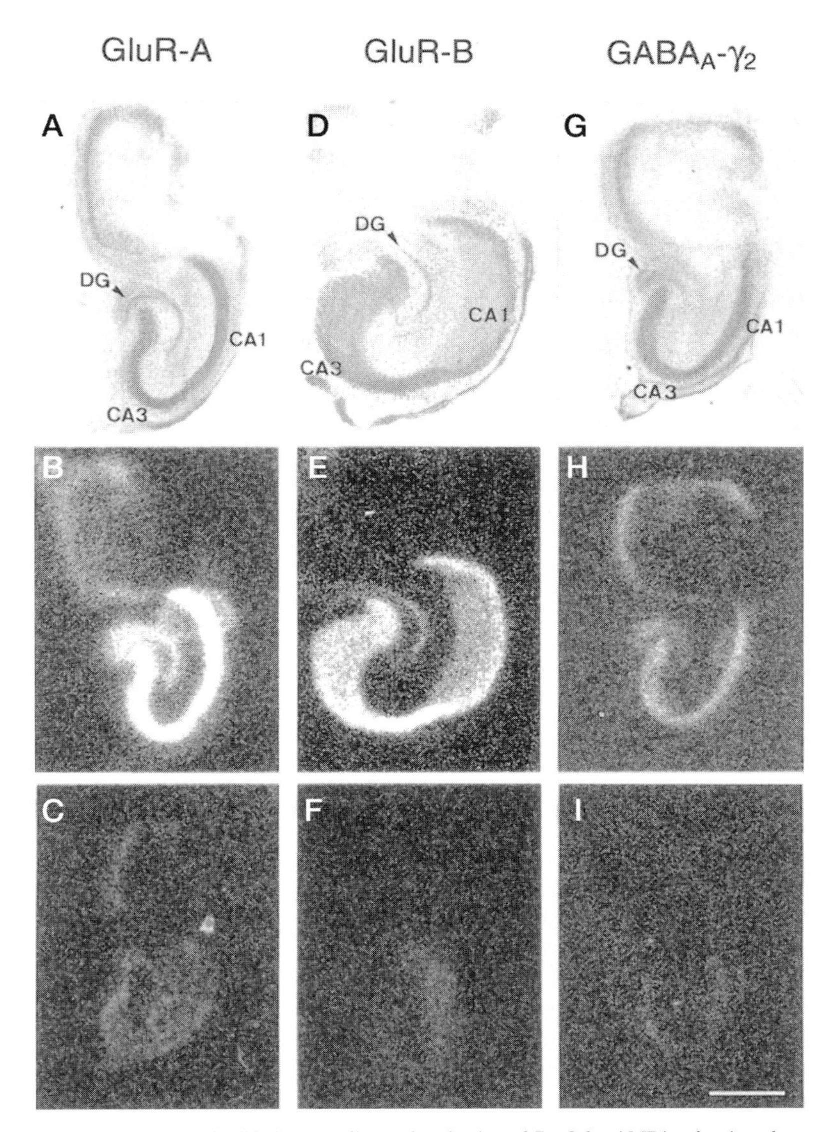

FIG. 8.1. Detection of mRNAs encoding subunits A and B of the AMPA-selective glutamate receptor and subunit γ_2 of the $GABA_A$ receptor in hippocampal slice culture preparations. Nissl stains of cultures hybridized with the labelled probe (A, D and G) demonstrate the morphological variability of the cultured hippocampi. The corresponding X-ray films (B, E, H) show the mRNA levels for each receptor subunit present in these cultures. The signals are displaced by adding excess unlabelled oligonucleotides to the probe as visualized by the X-ray films in C, F and I. Exposure time was 5 days for GluR-A and -B and 7 days for $GABA_A$-γ_2. The same magnification was used for all images. Scale bar in I = 1 mm. DG, dentate gyrus.

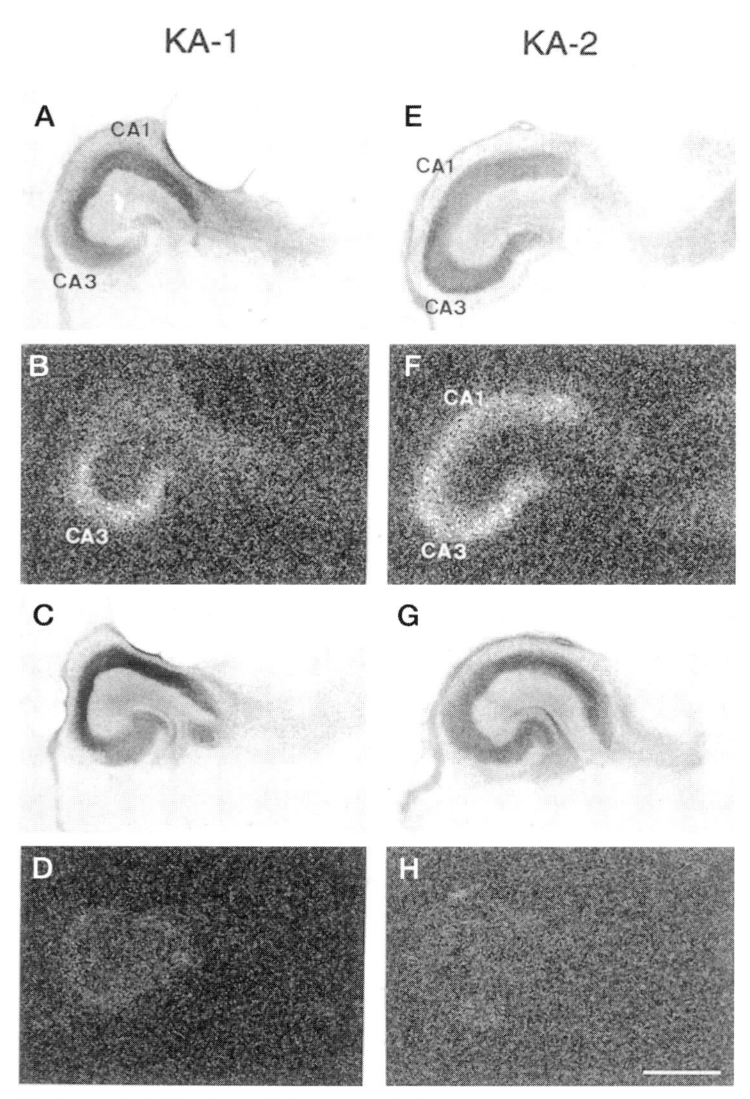

FIG. 8.2. *In situ* hybridization of hippocampal slice cultures to demonstrate the regionally selective expression of mRNA encoding the KA-1 subunit of the kainate-sensitive glutamate receptor. The signal on X-ray film obtained with the probe for KA-1 is restricted to the hippocampal CA3 region (B), whereas for KA-2, another subunit of this receptor family, mRNA levels are high in both the CA1 and the CA3 regions (F). Bright-field images show the Nissl stains corresponding to the cultures hybridized with probe (A, E) and to those hybridized with a mixture of probe and excess unlabelled oligonucleotides (C, G). The signal was displaced by addition of unlabelled probes as shown in the X-ray images of these control cultures (D, H). Scale bar in H = 1 mm.

(Fig. 8.1C,F,I), thus highlighting the specificity of binding for the labelled probes.

An example of the regionally selective gene expression in the culture system, which is very similar to observations *in vivo,* is shown in Fig. 8.2. Hybridization with a probe for subunit 1 of the kainate-sensitive glutamate receptor (KA-1) yielded an expression pattern that was mainly restricted to the CA3 region of the hippocampus (Fig. 8.2B). In contrast, labelling of both CA3 and CA1 (Fig. 8.2F) was obtained with a probe for subunit 2 (KA-2) of this receptor (Herb *et al.,* 1992). The weak labelling in the dentate gyrus, compared with *in vivo* expression patterns, probably results from the radiation-induced loss of granule cells (see Section 8.2). Bright-field images of the Nissl stain in Fig. 8.2C and G correspond to X-ray images of control cultures hybridized with excess unlabelled oligonucleotides (Fig. 8.2D and H).

In order to obtain images of the hybridization pattern at higher cellular resolution, hybridized slice cultures were coated with autoradiographic emulsion. Excellent cellular resolution can be obtained in regions where slices have flattened down to a single cell layer. For example, cerebellar slice cultures are especially well suited for emulsion autoradiography because tissue of cortical origin is almost always thinned to a monolayer (data not shown). The difference in resolution between exposure to film or emulsion can be appreciated in Fig. 8.3. mRNA expression in hippocampal cultures for GluR-A and GluR-C, two subunits of the AMPA-selective glutamate receptor family (Boulter *et al.,* 1990; Keinänen *et al.,* 1990), was investigated. Hybridized cultures were first exposed to X-ray film for 4 days and then to emulsion for 9 days with the results shown in Fig. 8.3A and D. The higher resolution obtained with emulsion is most conspicuous in the dentate gyrus. A portion of the signal in this area is much weaker or even missing (arrowheads) on X-ray film (Fig. 8.3B and E). Controls exposed to film are shown in Fig. 8.3C and F. Keinänen and co-workers (1990) have shown that mRNA levels for GluR-C within the hippocampal formation in sections from adult rat brain are highest in CA1 and dentate gyrus but diminish significantly in CA3. It can be seen from Fig. 8.3A and B that it is difficult to draw conclusions with regard to the relative amounts of mRNA between regions because the regional thickness of the cultures may not be the same. For semiquantitative analysis silver grains would have to be counted and normalized to the thickness of the cell layer or the number of cells.

The following protocol for ISH focuses on steps that have been adapted for investigations on organotypic slice cultures and thus differ from the detailed description in Chapter 1. For the use of non-radioactive cRNA probes on organotypic cultures, see Chapter 11 (Section 11.3.3). The use of alkaline phosphatase-conjugated oligonucleotides on organotypic cultures is described in Chapter 10 (Section 10.2.2).

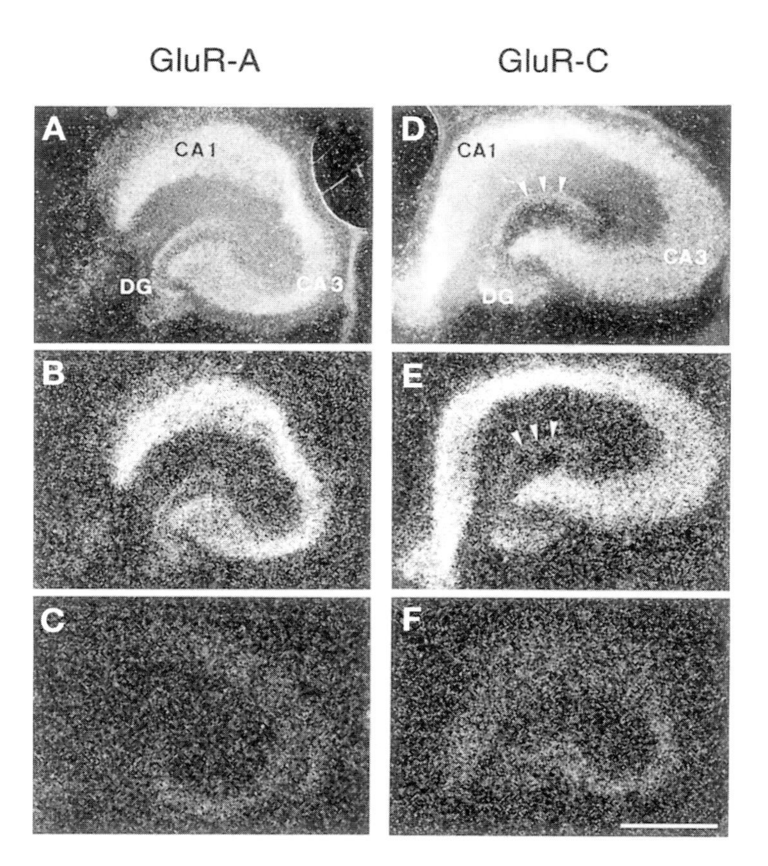

FIG. 8.3. Expression of mRNA in hippocampal slice cultures encoding the GluR-A and -C subunit of the AMPA-selective glutamate receptor, as revealed by autoradiography on photographic emulsion and X-ray film. For both probes, exposure of the hybridized cultures to emulsion resulted in images of higher resolution and sensitivity (A and D) than in those obtained from exposing the same cultures to X-ray film (B and E). Control cultures were incubated with a mixture of labelled and excess unlabelled oligonucleotides to demonstrate specificity by displacement of the X-ray signal (C and F). Exposure time was 9 days for emulsion and 4 days for X-ray film. Arrowheads indicate the signal in the dentate gyrus (DG), which was weaker when the culture was exposed to X-ray film than when it was exposed to emulsion. Scale bar in F 1 mm.

8.2 Preparation of Slice Cultures

Slice cultures of rat hippocampus are prepared and maintained *in vitro* as described in detail elsewhere (Gähwiler, 1984; Gähwiler *et al.*, 1991). Briefly, 5- to 7-day-old rat pups are killed by decapitation. The hippocampus is isolated, 425-μm-thick sections are cut on a tissue chopper under sterile conditions and exposed to ionizing radiation (650 rad, 4.5 min) in order to reduce the number of non-neuronal cells. It should be noted that, because some dentate gyrus cells are still dividing at the time of explantation, irradiation considerably reduces the size of the dentate gyrus in the cultures. The sections are mounted on glass coverslips (12 × 24 mm) in a droplet of clotted chicken plasma (Cocalico), placed in sealed culture tubes containing 0.75 ml of semisynthetic culture medium, and incubated on a slowly rotating roller drum at 36°C. This rotation serves the purpose of providing proper draining, feeding and aeration of the cultures, but also facilitates the gradual flattening and spreading of the slices. The medium consists of 25% horse serum, 50% Eagle's basal medium and 25% of either Hanks' or Earle's balanced salt solution (Gibco). After 17–20 days *in vitro* the slices become thinned to one to three cell layers and remain so for up to six weeks.

8.3 *In Situ* Hybridization

8.3.1 Fixation of Slice Cultures

Briefly, cultures are taken out of the culturing tube with curved, fine-pointed tweezers and placed in a Teflon rack (custom-made) holding 20 coverslips in vertical position, which has already been immersed in PBS. As it is not possible to mark the unfrosted coverslips, notes must be taken on the order of the cultures. The rack is transferred into 4% formaldehyde in PBS, freshly prepared from paraformaldehyde, and left for 5 min for fixation at room temperature. Cultures are washed three times for 5 min in PBS, dehydrated in 75% and 95% ethanol for 5 min each and air-dried at room temperature for about 45 min. Racks with cultures that are not to be used immediately for hybridization are placed in a staining trough or wrapped in aluminium foil and stored at –20°C until required. They can be stored in the freezer for up to a couple of weeks without any notable loss of quality in subsequent ISH experiments.

Long-term storage in 95% ethanol at 4°C should be avoided because the alcohol will cause the cultures to come off the coverslip during the washing procedure after the hybridization step, presumably as a result of some effect on the plasma clot that is used in the preparation of the slice cultures.

8.3.2 Hybridization

Frozen cultures are allowed to come to room temperature for 30 min. ^{35}S-labelled probes are diluted in "minimalist" hybridization buffer as described in Chapter 1 (Protocol 1.8). However, probes exceeding 300 000 dpm/µl are still used for the following reason: as determined by polyacrylamide gel electrophoresis (data not shown), a higher number of counts, obtained by liquid-scintillation counting, does not result from longer AMP tails (more than 50 nucleotides), but rather from a higher number of oligonucleotides carrying a radioactive tail. Probes above 300 000 dpm/µl are therefore diluted to around 200 000 dpm/µl with 1 × TE buffer. They are not found to produce non-specific binding.

Coverslips are placed horizontally and 30 µl of the hybridization mixture is placed onto each culture. This is sufficient to cover the whole hippocampal region and spreading of the mix is therefore not necessary. A glass slide (baked at 180°C for 3 h) is then gently lowered over the culture until the coverslip attaches to it by adhesion. The slide is inverted and placed horizontally with the culture now sitting on top of it. This way, up to four cultures with application of the same probe can be attached to one slide and its frosted end can be marked with pencil. There is usually no problem of air bubbles. To perform the hybridization, large slide-storage boxes (Kartell, Milano, Italy; 100 slides/box) containing tissues soaked with 10 ml of solution (4 × SSC/50% formamide) are used as humidifying chambers. Up to 14 slides (i.e. up to 56 cultures) can be placed horizontally per box. The boxes are sealed with electrical tape to maintain humidity and incubated at 42°C overnight in an oven.

8.3.3 Washing

Slides are removed from the humidifying chamber and lowered at an angle into a jar containing 2 × SSC at 50°C with the coverslips facing downwards. In this way, the coverslips slide off the slides easily and are prevented from landing culture side down on the bottom of the staining jar. The jar should be placed on a black surface to outline the delicate coverslips for easier handling. The cultures are transferred one-by-one back into a Teflon rack immersed in 1 × SSC at room temperature. The order of the hybridized cultures must be noted separately. The rack is transferred into pre-warmed 1 × SSC at 60°C and left for 20 min. A series of rinses at room temperature is then performed for 5 min each: 1 × SSC, 0.1 × SSC, 75% ethanol and 95% ethanol. Cultures are subsequently allowed to air-dry.

8.4 Autoradiography

8.4.1 Exposure to X-ray Film

Dry coverslips are attached to cardboard with tape and exposed to βiomax autoradiographic film for 3 days to two weeks depending on the probes used and the efficiency of their tailing. Autoradiographic ^{14}C-labelled microscale strips (Amersham Biosciences) are found to be very convenient for comparison of films exposed for different lengths of time (see Chapter 9, Section 9.4.1). After exposure, cultures are removed for subsequent counterstaining or exposure to emulsion by gently peeling the tape off.

8.4.2 Exposure to Autoradiographic Emulsion

Coating with Emulsion

Emulsion (NTB-2, Kodak) is melted in a waterbath at 42°C, and an aliquot is poured into a black film-roll container. A glass slide is dipped and checked for air bubbles. When bubbles are present, the emulsion is not touched for another 15 min. Cultures are dipped individually into the undiluted emulsion for about 5 s. White plastic forceps (metal must not be used!) were used for more convenient handling under safe-light illumination. Excess emulsion is allowed to drain by gently touching a tissue with the lower end of the coverslip; the coverslips are then placed 'on end' in a plastic rack to dry. They are left in complete darkness for 4 h overnight. Coated coverslips are transferred to Teflon racks with plastic forceps, then racks are placed into light-tight boxes containing a bag of silica gel and stored for exposure at 4°C for the required time (between 5 days and four weeks). See Protocol 1.9 (Chapter 1) for the use of Ilford K5 emulsion.

Developing and Counterstaining

Exposed boxes are allowed to come to room temperature for 1 h. Emulsion-coated coverslips are developed according to the manufacturer's instructions at 15°C. Briefly, cultures are placed in Dektol developer (Kodak; diluted 1 : 1 in water) for 2 min, rinsed in water for 30 s, fixed in a freshly prepared solution of 30% sodium thiosulfate for 10 min, and rinsed in running tap water for 20 min. For counterstaining, the rack is transferred to a solution of 0.2% Toluidine Blue and left for several minutes. Cultures are then rinsed for 1 min in water, destained in 0.01 M HCl until the emulsion layer is free of blue stain, rinsed in water for 1 min, destained in 70% ethanol until excess staining of the tissue has been removed (10–30 min, depending on the thickness of the cultures and on the time of staining). Cultures are dehydrated in 95% ethanol, twice in 100% ethanol and cleared twice in 100% xylene for 5 min each. Finally, coverslips are mounted culture side down on microscope slides by slowly lowering them on to a drop of mounting medium (Eukitt, Kindler, Germany).

Acknowledgements

We would like to thank Drs Scott Thompson and Peter Streit for critical comments on the manuscript. Special thanks are due to Dr Beat H. Gähwiler for providing the cultures and to Franziska Grogg and Lotty Rietschin for their excellent technical assistance in preparing them. This work was supported by the Swiss National Science Foundation (Grant nos. 31-28652.90 to P. Streit and 31-27641.89 to B.H. Gähwiler) and the Sandoz Research Institute in Berne, Switzerland.

References

Bauer, R., Hoflehner, J., Doblinger, A., Kapeller, I., and Laslop, A. (2000). *Eur. J. Neurosci.* **12,** 2746–2756.

Boulter, J., Hollmann, M., O'Shea-Greenfield, A., Hartley, M., Deneris, E., Maron, C., and Heinemann, S. (1990). *Science* **249,** 1033–1037.

Finsen, B.R., Tonder, N., Augood, S., and Zimmer, J. (1992). *Neuroscience* **47,** 105–113.

Gähwiler, B. H. (1984). *Neuroscience* **11,** 751–760.

Gähwiler, B. H., Enz, A., and Hefti, F. (1987). *Neurosci. Lett.* **75,** 6–10.

Gähwiler, B. H., Thompson, S. M., Audinat, E., and Robertson, R.T. (1991). *In* "Culturing Nerve Cells" (G. Banker and K. Goslin, Eds), pp. 379–411. MIT Press, Cambridge, MA.

Gähwiler, B. H., Capogna, M., Debanne, D., McKinney, R. A., and Thompson, S. M. (1997). *Trends Neurosci.* **20,** 471–477.

Gerfin-Moser, A., Grogg, F., Rietschin, L., Thompson, S. M., and Streit, P. (1995). *Neuroscience* **67,** 849–865.

Herb, A., Burnashev, N., Werner, P., Sakmann, B., Wisden, W., and Seeburg, P. H. (1992). *Neuron* **8,** 775–785.

Keinänen, K., Wisden, W., Sommer, B., Werner, P., Herb, A., Verdoorn, T. A., Sakmann, B., and Seeburg, P. H. (1990). *Science* **249,** 556–560.

Killisch, I., Dotti, C. G., Laurie, D. J., Lüddens, H., and Seeburg, P. H. (l991). *Neuron* **7,** 927–936.

Moser, A. M., Müller, M., Gähwiler, B. H., Thompson, S. M., and Streit, P. (1993). *Experientia* **49,** A74.

Müller, M., Gähwiler, B. H., Rietschin, L., and Thompson, S. M. (1993). *Proc. Natl Acad. Sci. USA* **90,** 257–261.

Werner, P., Voigt, M., Keinänen, K., Wisden, W., and Seeburg, P. H. (1991). *Nature* **351,** 742–744.

CHAPTER 9

QUANTITATIVE ANALYSIS OF *IN SITU* HYBRIDIZATION HISTOCHEMISTRY

A. L. Gundlach* and R. D. O'Shea**

*Howard Florey Institute of Experimental Physiology and Medicine and Department of Medicine, Austin and Repatriation Medical Centre,The University of Melbourne, Victoria, 310 Australia
**Department of Pharmacology, Monash University, Victoria, Australia

INTERNATIONAL REVIEW OF NEUROBIOLOGY, VOL. 47
ISSN/02 $00.00

9.1 Introduction

Past advances in several disciplines, including biochemistry and molecular biology, and the increasingly easy access to gene sequence information in publications and databases, culminating perhaps with the recent publication of the human genome sequence (IHGSC, 2001; Venter *et al.*, 2001), have continued to focus the attention of biological scientists on the study of known and novel genes. In the case of neuroscientists, the many relevant targets have included genes that code for neuropeptides and trophic factors, signalling enzymes, transmitter receptors, various mediators, and regulatory and structural proteins. This has created a corresponding increase in the need to understand the distribution and control of gene expression. Gene regulation may be studied using a large number of molecular biological techniques. Only *in situ* hybridization histochemistry (ISH), however, permits the *precise* regional and cellular localization of gene expression in a particular tissue, especially in morphologically and functionally complex tissues such as the nervous system (e.g. Young, 1990). Since changes in gene transcription, indicated by altered levels of mRNA encoding specific peptides or proteins, are physiologically relevant markers of cellular activity, the ability to perceive these changes and to determine their magnitude accurately is of great importance. When comparing levels of mRNA in different anatomical locations or developmental stages, or in response to pharmacological and physiological/surgical manipulations, or to pathology, quantitative analysis of ISH not only allows the identification of localized changes, but also measurement of their relative magnitude.

Many of the techniques used in the quantitative analysis of ISH are derived from those devised for neurotransmitter receptor autoradiography, the theoretical aspects of which have been reviewed elsewhere (e.g. Kuhar *et al.*, 1986; Davenport and Hall, 1988; Davenport and Nunez, 1990). An understanding of how X-ray films and nuclear emulsions respond to radioactivity is also useful in the design and analysis of ISH experiments and these subjects are excellently covered in the readable and informative text by Rogers (1979).

This chapter, which has undergone some revision for the new edition, still aims to review the theoretical considerations which should be observed to allow the accurate and meaningful quantification of ISH, and to provide practical advice and instructions for this analysis (see O'Shea and Gundlach, 1994). This chapter is focused on convenient computerized methods for the quantification of ISH results, as these are now very commonly used and are usually more efficient than manual methods. The methods of analysis described are most relevant to the results of ISH experiments obtained using radioisotopically labelled DNA oligonucleotides which hybridize to mRNA in thin tissue sections, although many of the theoretical aspects of this analysis are of relevance for a wider range of experimental situations.

9.2 General Considerations

9.2.1 Experimental Design

In designing ISH experiments with a view to quantifying results, it is intuitive that the researcher should attempt to eradicate or at least minimize any potential sources of variation between the treatment of different experimental tissue sections. While identical processing of sections may not be possible in all instances, samples to be analysed concurrently (i.e. those to be directly compared with each other) should ideally be prepared at the same time and processed together, since between-experiment variation contributes most significantly to the overall variability of results (e.g. McCabe *et al.*, 1993). For accurate quantification, sources of variation at each step in the preparation and hybridization of sections and in the exposure and subsequent analysis of images should be kept in mind (Table 9.1).

9.2.2 Choice of Radioisotope

Although many radioisotopes (^{3}H, ^{14}C, ^{32}P, ^{33}P, ^{35}S and ^{125}I) have been used in ISH, a number of factors determine the suitability of different isotopes for a particular application. The anatomical resolution provided by an isotope is inversely related to the maximum energy of the β-particles it emits (Rogers, 1979). Probes with low specific activities and long half-lives (^{3}H and ^{14}C) provide excellent cellular resolution (e.g. Harlan *et al.*, 1987 for ^{3}H) but require long exposure times, and for this reason are rarely used in ISH. One problem unique to ^{3}H is its differential self-absorption by tissues of different lipid content, leading to apparent variations in signal density where none exist. However, delipidating sections prior to exposing them to film or emulsion can solve this problem. ^{32}P-labelled probes, which emit β-particles of very high energy, provide rapid results at the cost of cellular resolution, and may be used in situations where a preliminary result, or regional rather than cellular resolution, is required. The short half-life of this isotope (14.3 days) also gives it a relatively limited "shelf-life". Oligonucleotide probes labelled with ^{35}S and ^{33}P both provide good cellular resolution without requiring excessively long exposure times, and these are most commonly used in ISH. For these higher energy isotopes, section thickness and the thickness of the emulsion layer are both critical in producing consistent results, since β-particles emitted from the decay of these isotopes from the entire thickness of the section (normal range 8–20 μm) can produce a latent image on silver halide crystals throughout the extent of the emulsion layer (Rogers, 1979).

TABLE 9.1
STRATEGIES TO MINIMIZE VARIATION IN QUANTITATIVE ISH RESULTS.

Procedure	*Control*
Obtaining tissue	Sacrifice animals from different experimental groups at same time of day and in random order using an identical routine Identical dissection and freezing of tissues[a]
Cutting sections[b]	Cut sections from as many different groups as possible onto same microscope slide[c] Position sections cut at the same anatomical level on the same slide[d] Randomize or alternate order of cutting and mounting sections from different groups on slide e.g. slide series 1: animal group A, B, C slide series 2: animal group B, C, A slide series 3: animal group C, A, B[e] (Consistent section thickness is critical)[f]
Section fixation, storage[g]	Fix all sections for a particular experiment at the same time and store together
Application of probe	Use the same batch of labelled probe for all sections to be analysed together[h]
Hybridization, washing	Perform at same time for all sections
Application to film, developing film	Co-expose all sections at same time and for same period (expose more than once if needed; see Section 9.4.8 of text)[i] Position slides from each series throughout area of film (to minimize effects of variations in emulsion thickness) Only use X-ray cassettes that are in good condition (leaking or bent cassettes will result in uneven exposure or apposition of slides to film) Standardize film-developing procedure[j]
Emulsion dipping and development	Process all sections together using same batch of emulsion Standardize dipping procedure (temperature, time taken to dip each slide, drying)[k] Use same batch of emulsion Standardize emulsion development procedure

[a] For brain tissue, rapid removal of the brain at room temperature, followed by dissection in the coronal or sagittal plane with a single-edged razor blade is recommended. The brain is then rapidly frozen in a plastic weighing boat, with cut surface down, on (not in) liquid N_2. For dissection of the rat brain for sectioning in the flat-skull position (corresponding to the atlas of Paxinos and Watson, 1986), the Watson–Paxinos Stereotaxic Brain Blocker (David Kopf Instruments, Tujunga, USA) may be used.

[b] Sections can be collected onto slides stored in the cryostat, and thawed against the forefinger or the back of the hand. This allows section flattening and avoids wrinkling and folding of sections.

[c] For rat forebrain or similar sized tissues, 3–4 sections per slide is optimal; for rat medulla oblongata or mouse brain, up to 8–10 sections may be placed on each slide. Sections

should be positioned centrally on the slide to avoid uneven hybridization conditions near the edge of the parafilm coverslip and uneven emulsion thickness near the edge of the slide. If structures of interest (e.g. brain nuclei) are located laterally in the section, single hemispheres from control and experimental animals may be carefully aligned prior to freezing and sectioning (see e.g. Bhat *et al.,* 1992). This saves time and space, and gives directly comparable thickness of sections.

[d] When cutting rat brain, anatomical level may be monitored using a low-power stereomicroscope with a dark-field or trans-illumination base. While the section is still moist, anatomical details of white matter tracts versus grey matter-enriched areas can be clearly identified and related to the acetylcholinesterase-stained sections of the Paxinos and Watson (1986) atlas. While it is not always possible or *essential* to have closely matched anatomical levels on each slide, it makes analysis more convenient and ensures that sections that are analysed together have been processed identically.

[e] This controls for any effects of exposure of the sections to different periods at room or cryostat temperature prior to fixation or other pretreatments and storage before use.

[f] Uneven section thickness results in variable film OD with higher energy radioisotopes. Hence the usual precautions regarding the effectiveness of the cryostat/microtome should be adopted to ensure that the machine is not cutting uneven, 'thick–thin', shattered or wrinkled sections. Machines with fully adjustable mounting for tissue chucks allow more flexibility in collecting anatomically aligned sections.

[g] Post-sectioning fixation is optional under some common procedures (see Dagerlind *et al.,* 1992; A. L. Gundlach laboratory articles) but if tissue is to be fixed, time and conditions should be standardized. The ability to carry out ISH without any fixation may remove inconsistencies of fixation, particularly after perfusion fixation, and may in fact give higher specific/background hybridization ratios.

[h] Labelling of probes to an identical specific activity is extremely difficult, so it is best to hybridize all experimental sections to be analysed for a single experiment with the same batch of probe or mixture of probes.

[i] There will be occasions when an increase or decrease in mRNA levels will mean that optimization of exposure time for control or treated sections will result in under exposure or saturation of other related images. Film or emulsion saturation (OD greater than 1.0 or silver grains overlying each other) seriously affects the accuracy of sampling, and should be avoided if at all possible (Rogers, 1979; Davenport and Hall, 1988; Davenport and Nunez, 1990).

[j] Many labs use automated developing machines for Kodak BioMax film. If other types of films requiring separate/manual development are used, it is important to use the recommended development temperature, time and chemicals on *all* occasions.

[k] This is necessary for obtaining emulsion layers of consistent thickness, which are vital for accurate quantification (see Section 9.5.3 of text). A potential problem with emulsion drying is that some emulsion will run from the top to the bottom of the slide before drying. For this reason avoid placing sections too close to the extreme ends of the microscope slide surface.

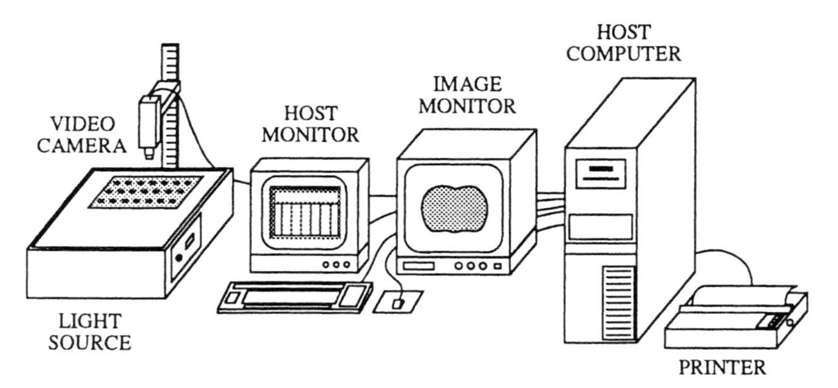

FIG. 9.1. Basic components of a computerized image analysis system (see text for details).

9.3 Computerized Image Analysis Systems

The purpose of a computerized image analysis system (IAS) in the quantification of ISH is to rapidly, accurately and conveniently convert visual experimental results into a format that can be processed to provide meaningful quantitative or semiquantitative information. A variety of approaches may be taken to achieve this data-processing objective, and this is reflected in the variety/number of IASs available to the researcher. In a more significant way than at the time of writing for the first edition (O'Shea and Gundlach, 1994), advances in digital technology and a better understanding by manufacturers and suppliers of the needs of bioscience research, means that modern IASs are faster, more powerful, easier to operate and less expensive than earlier systems. Many now provide highly specialized procedures for rapid and accurate analysis of visual data. While many of the earliest systems were essentially custom-built and programmed, most modern microcomputer-based IASs may be purchased complete ('off the shelf') or assembled around, and effectively used with, standard computer hardware.

The basic components of such a system, which provide a means of image acquisition, processing, display, analysis, storage and output, are illustrated in Fig. 9.1. The input device used depends on the type of visual data being analysed and the degree of detail required, and is generally a monochrome or colour video camera attached to a suitable lens or microscope. Specific types of input devices will be discussed in more detail in sections dealing with quantification of different types of ISH data.

Light intensities from the image are detected by the video camera as analogue electrical signals, which are converted to digital format for processing by the computer. This digitization step breaks the image into discrete spatial elements (pixels) that each possess values for density (in grey levels) and specific X–Y location. Density is generally recorded with 8 or 12 bits of precision (corresponding to 256 or 4096 grey levels), and 256×256 to 2048×2048 pixels spatial resolution, depending on the amount of space allocated to image memory. True colour systems generally assign 24 bits to colour, 8 bits each to red, green and blue. Greater memory space allows more detail to be obtained from images, and image memory is now relatively inexpensive, as are the more powerful processors required to manipulate these larger amounts of data. A resolution of 512×512 pixel $\times$ 256 grey appears to be adequate for handling film images of ISH, and is also suitable for analysis of emulsion images.

In choosing an IAS, the researcher is faced with a range of options. Most modern systems are capable of analysing results from a variety of applications, and, in the absence of any special requirements, more than one system is likely to be suitable for the routine analysis of ISH results. Choice of systems must then be based on criteria other than just the ability to perform the necessary quantification of experimental results. It may be difficult for the researcher to properly evaluate the performance of different IASs even if access to working systems is possible. Discussion with experienced users of different systems is invaluable, and may inform potential buyers of the ease of use as well as the advantages and shortcomings of particular systems. A study of the literature in an appropriate field of research may give some indication of which systems are commonly used, but will not necessarily reveal the best, most recent or most economical options.

The rapidity of change in computer technology leads to continual refinements in hardware *and* software, so the ability to upgrade or update components of the IAS easily is a distinct advantage. Systems with modular design and software update availability allow the researcher to keep pace with advances in image analysis technology, and lessen the risk of being left with obsolete equipment. Flexibility of hardware and software options within a given system also allows tailoring of the system to specific or multiple needs the user may have which were not present (or not envisaged) when the system was initially purchased, avoiding the expense of duplicating system components. Cost, or at least perceived value for money, is also likely to be an important consideration for most groups in selecting an IAS. Nowadays, several commercial suppliers do offer hardware and software upgrades, which prevents the IAS becoming obsolete and avoids the requirement for full replacement. Software packages are also available from publicly funded sources such as the NIH (e.g. NIH-Image program).

Our laboratory has always used a “microcomputer imaging device”

(MCID, M1 or M2) system from Imaging Research Inc. (St. Catharines, Canada). The current software runs on an IBM-compatible computer under the powerful Windows NT operating system. The MCID system is of modular design, and includes integrated software modules for many bioscience applications, some of which will be discussed in later sections. Custom software for specialist applications can also be obtained from the supplier. An advantage of microcomputer-based IASs is the ability to export data directly to spreadsheet and statistical packages (running on the same computer or another compatible machine) for further analysis, avoiding the time-consuming process of re-entering data for this purpose.

Different IASs use different methods of displaying images and other data. The MCID systems use two monitors: one to display measured data and interactions with the host computer (host monitor) and another to display images (image monitor; see Fig. 9.1). Displayed images may be processed in a number of ways, and various types of image data can be recorded. As well as visual output, numerical data and images may be printed (using a suitable high-quality printer for images) or stored on disk or CD for later retrieval. It is popular these days to import images into image-handling programs such as Adobe Photoshop and/or drawing/graphics programs, such as CoralDraw, to collate, manipulate and label images for publication or display (see also Chapter 1).

While not essential for the quantification of film ISH, visual enhancement of images is useful for illustrating experimental observations, as well as assisting in such processes as setting threshold densities during analysis. All modern computerized IASs allow for pseudocolour coding of grey-scale values. This feature can simplify the perception of differences in signal density, as the human eye can more readily distinguish differences in colour than in shades of grey, although in monochrome images, anatomical borders of brain regions are often clearer.

9.4 Quantification of Film Images of ISH

9.4.1 Introduction

The apposition of sections hybridized with radioactively labelled probes to X-ray film (Chapter 1, Section 1.9.2) allows the rapid quantification of gene expression at the regional (and to a degree cellular) level of brain and other tissues. The optical density (OD) of film images produced by these tissue sections is related (though not necessarily in a linear manner) to the regional radioactivity concentration, and thus represents the relative amount of target mRNA present in the sections. In order to determine regional radioactivity values (and hence relative abundance of mRNA), experimental images are best compared with those produced by samples of known radioactive content which are co-exposed to film with the hybridized sections. The production of radioactive standards, the fitting of standard curves and the sampling of experimental results (including extrapolation from standard curves) are all procedures that need to be understood in order to gain accurate and reproducible data. These latter aspects are generally well explained in the user manual of the IAS or relevant software. Some knowledge of the limitations of film-based autoradiography and its quantification is also important.

9.4.2 Resolution of Film Images

The degree of resolution (and hence the amount of anatomical information) which can be gained from film-based emulsions is limited in part by the emulsion type used. (Resolution is also influenced by the choice of radioisotope, section thickness, the closeness of apposition between the sections and film, and the length of exposure; these factors are discussed elsewhere in this chapter.) Different X-ray films are distinguished by the thickness and characteristics of their emulsion layer (the size of the silver halide crystals, the size of the developed silver grains and the sensitivity of the emulsion to β-particles of different energies). All of these factors determine the suitability of the film to a particular application as well as the resolution which can be achieved using the film (see Rogers, 1979). Many currently available X-ray films are suitable for use in ISH, with the most commonly used probably Kodak BioMax, which is made with a high silver content emulsion on one side of a clear plastic base, and produces images of good resolution from the shortest possible exposure time.

9.4.3 System Requirements for Quantification of Film Images

For accurate quantification of film images of ISH, sources of variation in illumination and recording of films must be removed or their effects minimized. Most commonly, a video camera attached to a photographic lens is used to capture images of the film, which is evenly illuminated from below. Any variation in camera performance or lighting conditions ultimately affects the accuracy of data recorded, so the use of optimal and standardized conditions for the digitization of images is essential for meaningful quantification. The major sources of error in image digitization conditions (variations in camera performance and in lighting parameters) may be controlled by careful selection of imaging components and attention to analysis conditions. The ability of the light source to produce stable illumination in the presence of variations in power supply and over sustained periods is critical for accurate measurement. A good IAS should allow for the correction of shading errors brought about by an uneven pattern of illumination or irregularities in camera performance (e.g. lower sensitivity at extremes compared with the centre of the field). These features, plus the uniformity of ambient lighting in the immediate environment around the IAS, assist the researcher/operator in obtaining accurate and reproducible results.

For analysis of film images the standard MCID M2 system uses a Sony DXC-930P CCD video camera (or similar) attached to a Nikon micro Nikkor 55-mm, f 2.8 lens. For additional image magnification a Vivitar extension tube set is used. Coupled to a Kaiser RS1 copystand, this configuration allows for a wide range of magnifications, appropriate for many different experimental situations (e.g. visualization of a whole rat brain section or analysis of detail in small brain nuclei). Film illumination is supplied by a Northern Light Precision Illuminator model B90 (from Imaging Research Inc.), which provides constant lighting (the intensity of which can be manually set) with digital readout of lamp power, and is specified to exhibit less than 0.05% drift over 12 h. External light in the image analysis environment may be controlled by dim ambient lighting conditions, although if preferred cardboard tubes can be used to surround the camera lens and extend down over the film to exclude room light.

9.4.4 System Calibration

The performance of any computerized IAS depends on how well the system is prepared to carry out the quantification required. For obtaining reliable measures, a standard quantification procedure should be adhered to, with analysis conditions being optimized for the type of analysis being performed. The precise details of calibration depend on the system being used and the particular application, but the basic processes involved are similar.

Camera Magnification and Focus

The camera position and focus must be set to provide appropriate magnification for the regions to be studied. It is obviously convenient to be able to digitize large regions in one step, so too high a magnification factor is to be avoided. For acceptable accuracy at the other extreme, magnification must be sufficient for regions to be at least three to four pixels wide (assuming 512 × 512 pixel image memory; Ramm, 1990). When targets of different sizes are to be analysed on a single film, it is more convenient if all sampling can be carried out at one magnification, even if different regions must be digitized separately.

Spatial Distance Calibration

If absolute distance or area measurements are to be recorded from film images, distance calibration must be performed each time the system magnification is altered. A good-quality transparent perspex ruler can be placed under the video camera/lens to define distances (in millimeters or micrometres), which are then used to convert the number of pixels to an absolute measure of distance or area. In the absence of this calibration, the IAS can still record distances or areas in pixels, but these measurements are only relevant to the particular magnification conditions under which they are recorded.

Shading Error Correction and Lighting Adjustment

Shading error correction is a pixel-by-pixel adjustment process that is performed to compensate for variations in background density (shading) over the field of view. These variations are due to inconsistencies in the illumination pattern (from the light source and the imaging environment) and in camera and lens performance. Shading error correction needs to be carried out when the relationship of the imaging components changes (e.g. due to moving the light source or changing the magnification). Once this correction is performed, data may be sampled from multiple films as long as the imaging components are not moved relative to each other. For the MCID M2 system, shading error is corrected by digitizing a blank field with the lighting set so that the OD of this blank field is at the mid-range of values encountered in analysis.

Reading regional density data at the extremes of the OD range must be avoided for accurate quantification of film images. Regulating film exposure times can prevent saturation of images (see Section 9.4.8), while correct adjustment of the illumination level for analysis avoids the problem of images being too bright. Optimal illumination levels may vary for different IASs, and for analysis with MCID systems it is recommended that lighting is adjusted to produce OD values of 0.05–0.1 for areas of film background.

Standard Curve Fitting

Film response to radioactive emissions is not linear, so a way of relating film OD to tissue radioactivity content is required. To achieve this, film images produced by a set of standards (see Protocol 9.1) of known radioactivity are analysed, and a calibration curve is interpolated from these standards to assign values to experimental samples between the points on this curve. Thorough illustrative and descriptive details of how to do this are provided in IAS user manuals. Owing to slight differences in film apposition and between-film variation in silver content (however small these factors may be), it is good practice to co-expose radioactive standards with every sheet of film and to create a standard curve for each film exposure. Each standard may then be digitized, and OD (± area; Protocol 9.1) recorded. Once this is done for all standards over an appropriate range for the film, a curve is fitted to the density values (calibrated in appropriate units; Protocol 9.1). A number of curve types have been used to fit radioactivity standards, with log–log plots most commonly used. The 'best' curve type to use with a given set of standards and exposure conditions is the one which provides the most accurate fit to the points, and IASs generally provide a range of mathematical functions from which to choose, as well as advice on their use. It is not necessary to always use the same curve type as different exposure conditions and different sets of standards can produce different responses. It is possible to record uncalibrated OD measurements (i.e. without reference to a standard curve), to ascertain whether one target is darker or lighter than another. This procedure is, however, critically dependent on the lighting conditions used, and values recorded in this way have no fixed reference point.

The use of radioisotopes with a relatively short half-life (e.g. ^{33}P, ^{35}S) necessitates the preparation of new standards on a fairly regular basis to ensure that standards produce a range of OD values that is appropriate for comparison with experimental samples. One strategy for avoiding this procedure is to cross-calibrate standards of the isotope used in ISH with standards of an isotope which emits β-particles of similar energy but which has a longer half-life. ^{14}C fulfils these criteria, and standards incorporating this isotope may be purchased or prepared using any stable ^{14}C-labelled compound and a procedure similar to that of Protocol 9.1. A conversion factor, relating the film OD produced by the two radioisotopes, can be determined from co-exposure of both sets of standards to film. In subsequent studies, ^{14}C standards alone can then be co-exposed with ISH slides and the conversion factor used to relate film OD to tissue concentration of the isotope used to label ISH probes (see Miller, 1991; Eakin *et al.*, 1994).

Units of radioactivity used in creating standard curves are not critical, and there are many similar ways of describing the same values. Thus, while we and others normally prepare our standards and calibrate the IAS in dpm/μg wet weight or per mg protein, units such as nCi/mg protein or fmol/μm^2 are equally valid. However, area-based units and measurements may be more appropriate in situations where the physical size of labelled cells/nuclei are altered in an experimental situation, relative to control (e.g. in the

PROTOCOL 9.1 PREPARATION OF ^{35}S-LABELLED BRAIN PASTE STANDARDS

Many investigators have adopted the use of isotope-impregnated tissue standards as a means of calibrating the IAS for quantitative analysis. While a large number of protocols exist for the preparation of these standards, most of these are essentially modifications of a similar procedure. The following method is simple and convenient and provides reproducible results over many batches of radioactivity.

For the preparation of 10 standards that span a 1000-fold concentration range, the following steps may be used:

1. Homogenize whole brain (or brain less cerebellum) from two rats with mortar and pestle. Place ~100 μl aliquots in separate 1.5-ml microcentrifuge tubes,[a] labelled 1–10. Centrifuge briefly (30 s) to remove air bubbles and pack down homogenate.
2. Prepare dilutions of radioactivity in distilled H_2O (dH_2O). Add 1 μl of [^{35}S-]dATP (specific activity 12.5 Ci/μl or ~2.8×10^7 dpm/μl) to 19 μl of dH_2O to make dilution "A". Count 1 μl (or convenient volume) of "A" (should be about 1.4×10^6 dpm/μl).
3. For 50 000 dpm/mg wet weight brain tissue (or 5 million dpm/100 mg), use $(5 \times 10^6)/(1.4 \times 10^6)$ or 3.6 μl of "A" in 100 mg brain paste → standard 1.
4. For 10 standards distributed over a 1000-fold range (50–50 000 dpm/mg) each successive dilution needs to be by a factor of $1000^{1/9}$ or 1/2.15. So:
   Add 10 μl of "A" to 11.5 μl dH_2O → dilution "B".
   Then add 3.6 μl of "B" to 100 mg brain paste → standard 2.
   Repeat for dilutions "C" to "J" (standards 3–10).
5. Mix all standards well and centrifuge briefly. Repeat the "mix/centrifuge" step 3 times.
6. Count radioactivity from about 10 mg (record weights) of each brain paste standard to get dpm/mg values. This is best done by adding tissue to preweighed scintillation vials, recording the final weight, and solubilizing radioactivity with scintillant overnight before counting. (Some researchers prefer to count radioactivity from a number of cryostat sections, cut as in step 8 below, either weighing the sections or simply recording an average amount of radioactivity for sections of each standard, which can be related to the area of the section during system calibration.)
7. Mix again, centrifuge and freeze standards.
8. Cut sections of desired thickness (same as thickness to be used for experimental tissues). Two standards can be conveniently mounted on the chuck at the same time and cut concurrently. To remove frozen standard blocks from microcentrifuge tubes for cutting, cut a small piece from the bottom of the tube and carefully shake or push block out.

[a] An alternative to the use of microcentrifuge tubes for the preparation of brain paste standards is to use 1-ml disposable plastic syringes, giving standards of identical

diameter, which may be removed by cutting off the tip of the syringe and squeezing out the frozen standard blocks.

[b] In our experience, for studying transcripts that are extremely abundant or very rare (e.g. vasopressin mRNA in magnocellular neurons or mRNAs encoding neurotrophic factors such as GDNF, respectively) the concentration of radioactivity in the standards may have to be increased or reduced, respectively, beyond this nominated range.

---

hypothalamus where magnocellular neurons undergo significant changes in volume and gene expression with alterations in osmotic balance, Burazin *et al.*, 2001; Zhang *et al.*, 2001).

9.4.5 Sampling of Film Images

IASs generally provide a choice of sampling methods for the quantification of film images. These sampling methods range from manual (where the user traces the outline of areas of interest) to fully automated (where all areas darker than a particular threshold OD or grey-scale value are sampled). The choice of method is dependent on such factors as the type of targets being studied (e.g. discrete brain nuclei with "patchy" labelling of neuron mRNA, or a relatively homogeneous pattern produced by dendritic or glial mRNA labelling. For example, see Figs 1.13 and 1.14 in Chapter 1), and the type of quantitative data required (e.g. density, target area, or number of discrete targets). An ongoing, casual survey of the literature dealing with ISH reveals no real consensus of sampling methods or form of reported data. Rather, individual researchers or groups have adopted their own routine quantification procedures. Thus, it suggests that no single method will work well for every application and in deciding which method(s) to use, the twin objectives of accuracy and reproducibility must be foremost in mind. The most common sampling methods, and some of their potential applications, are discussed below.

Manual Outlining

This is the most time-consuming method of sampling data, and may also be subject to the greatest operator bias and between-operator variations in sampling (e.g. see Eilbert *et al.*, 1990). For some applications this method may still be the most suitable or the only method available, although more recent image analysis software tends to provide faster and more reproducible alternatives. In its favour, manual or "semi-manual" sampling allows the skilled operator to define and analyse complex shapes or nuclei that vary in shape and size at different anatomical levels and which are thus

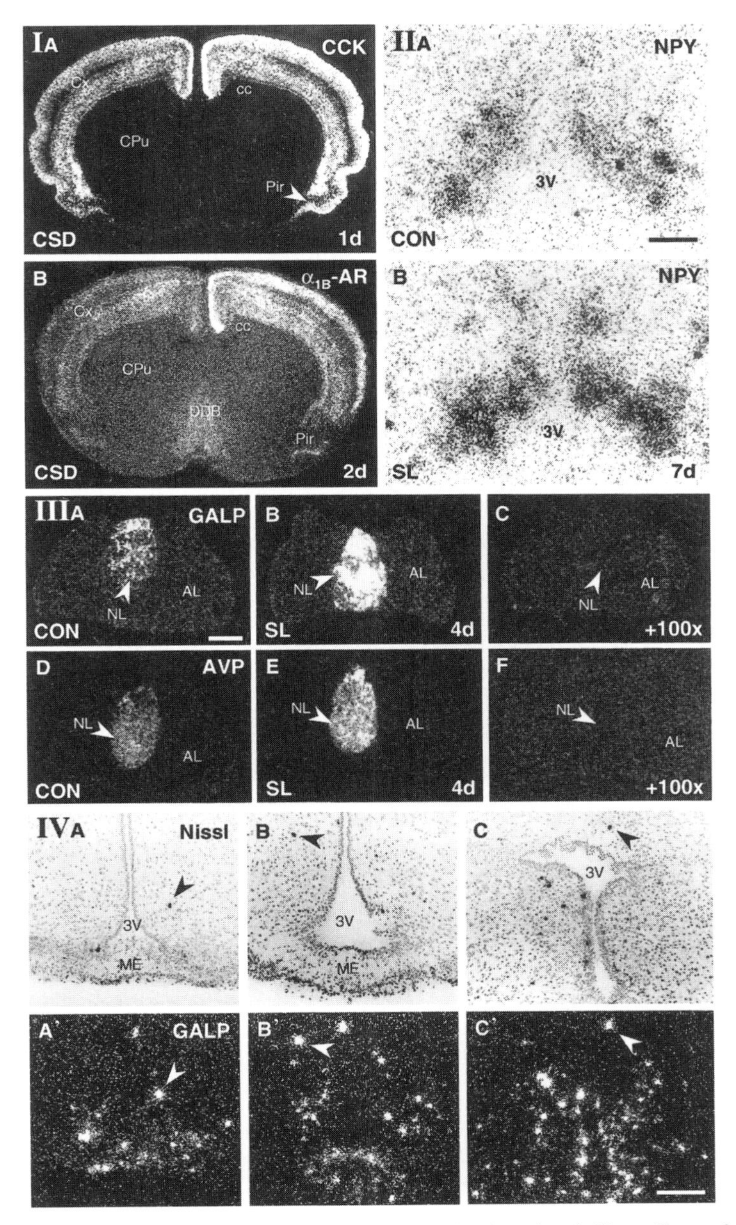

FIG. 9.2. Examples of different types of *in situ* hybridization signal (X-ray film and nuclear emulsion) that can be studied using the quantitative methods described in this chapter and which suit the different methods of data sampling and reporting (see text for further details).* I. Increases in (A) cholecystokinin (CCK) and (B) α_{1B}-adrenergic receptor (α_{1B}-AR) mRNA

difficult to outline using rigid templates. For example our laboratory has employed this approach to quantify alterations in levels of gene expression in different cortical layers following unilateral cortical spreading depression (Fig. 9.2I; Shen and Gundlach, 1998, 1999). Density and area of hybridization recorded using this method both provide meaningful data.

Geometric Sampling Tools

Most IASs provide a number of these tools, which the user may alter in size and orientation. The operator then simply positions the shape over the regions of interest to record signal density and other data. This method is relatively fast, and allows areas of identical size and shape, and of appropriate dimensions for the investigated regions to be sampled from multiple sections or samples. Some operator decision is required in the choice of size and shape of the sampling window and in its placement, but consistency of analysis may be achieved by always positioning the window in the same way (e.g. over the most lateral aspect of a nucleus or the area of densest signal).

FIG. 9.2 *continued*

throughout the ipsilateral cerebral cortex at different times (1–2 days) following an acute, unilateral episode of cortical spreading depression (CSD). This data was analysed using manual sampling of different cortical layers in X-ray film images (see Shen and Gundlach, 1998; unpublished observations). II. Neuropeptide Y (NPY) mRNA levels in the arcuate nucleus under (A) control conditions and (B) after salt-loading (2% NaCl in drinking water; SL) for 7 days. In this case, the area occupied by the hybridization signal was increased and this was assessed and reported (see O'Shea and Gundlach, 1995). III. Increases in (A,B) galanin-like peptide (GALP) and (D,E) arginine vasopressin (AVP) mRNA in the neural lobe of the rat pituitary gland produced by salt-loading (SL) for 4 days. (C,F) Negligible levels of non-specific signal observed in the presence of a 100-fold excess of each respective unlabelled probe(s) (+100×). Automatic sampling of film images was used in these studies to assess mean signal density and hybridization signal area. Manual measurements were also made of neural lobe area per section, to allow the calculation of a proportional hybridization area (p.a.) for each section. Subsequently, a density × p.a. calculation was used to reflect the relative amount of hybridization (see Shen *et al.*, 2001). IV. (A–C) Bright-field and (A'–C') dark-field images illustrating that GALP mRNA in rat arcuate nucleus is only present in a small number of cells at different rostrocaudal levels of the nucleus (see Larm and Gundlach, 2000). Arrowheads indicate examples of heavily labelled neurons. Such scarce and diffuse distributions are better suited to manual or automated analysis at the nuclear emulsion level, which allows measurement of numbers of labelled cells and density of grains per cell. AL, anterior lobe; cc, corpus callosum; CPu, caudate putamen; DDB, dorsal diagonal band; ME, median eminence; NL, neural lobe; Pir, piriform cortex; 3V, third ventricle. Scale bars 150 μm (II), 500 μm (III), 250 μm (IV).

**Note:* Images displayed are not necessarily representative of those used for the actual analysis, but instead may represent longer exposure images, which better portray hybridization signal photographically.

This method of sampling is only useful in recording density data, and would not be appropriate in situations where the size of the target under study is subject to variation, due to experimental treatments (for instance, if cells from a larger area express a particular transcript following treatment; e.g. Fig. 9.2II; O'Shea and Gundlach, 1995).

User-defined Templates

User-defined templates are used in the same way as the geometric sampling tools, and may be saved and recalled for later use. These templates allow consistent sampling of density values from regions of interest that are not accurately defined by geometric sampling tools.

Automatic Sampling

Using automatic sampling methods, outlines are created by the image analysis software using OD or grey-scale thresholds to distinguish targets from background areas. The operator must define a sample window (by selecting the whole of a section or only some part of it) and select the threshold level. This threshold may be fixed for the sampling of all areas or may be varied by the operator to compensate for changes in background signal. Determination of the threshold level introduces the possibility of inappropriate sampling for some targets if the threshold level is fixed at a particular OD or grey-scale level, or of operator bias or inconsistency when variable thresholds are used. Care must also be taken to accurately define an appropriate sampling window. Automatic sampling may be of use when regions of study are of variable size, but is difficult to perform when targets are poorly distinguishable from background. Selection of appropriate threshold levels is critical for accurate and reproducible automatic sampling, as small changes in threshold values may lead to large variations in average density or target area. (A recent example of automatic sampling of hybridization density and area data, combined with subsequent manual tissue area sampling on counterstained sections is described in Shen *et al.*, 2001; see Fig. 9.2III).

Other Sampling Strategies

Whichever sampling methods are employed, the operator must also decide whether and how to distinguish signals of interest from background and how to assess "specific" hybridization signals. For this reason, the techniques of redirected sampling and image subtraction that are common to many modern IASs deserve some attention at this point.

Using *redirected sampling*, multiple images may be overlaid and stored in different channels or different image memory areas. Redirected sampling may be used to define regions of interest (using histological identification by appropriate staining of sections) or to define "non-specific hybridization" (using consecutive sections to those showing "total" hybridization). In the first of these cases, the autoradiographic image of hybridization signal is digitized in one channel, and a corresponding section, stained with a histochemical or biochemical marker to identify regions of interest, is aligned with the first and stored in another channel, linked to the original one. Targets are defined from the histology (using outlining or geometric sampling tools), and data are read from the corresponding area of the autoradiograph. In this way, data may be obtained from cell groups/nuclei that are not easily identified by their hybridization pattern alone. Another possibility, or an extension of the above method, is to digitize an aligned image of "non-specific hybridization" into an additional channel, so that values of total and non-specific hybridization may be recorded from the two channels simultaneously. From a practical standpoint, this may be considered a relatively expensive long-term option, as large amounts of cold oligonucleotide probe(s) are needed to generate images of "non-specific hybridization". The issue of controls for specific hybridization is discussed further in Section 9.4.7.

Redirected sampling can reduce uncertainty in defining anatomical targets of interest and assist in sampling appropriate areas from multiple sections simultaneously. If redirected sampling is not available on the IAS being used, it may be possible to simulate this feature. Outlines may be drawn to highlight areas of interest on a digitized image of a stained section, and then, while retaining the overlaid outlines, other sections may be digitized and aligned with these outlines and data sampled from the appropriate areas.

Image subtraction is another tool commonly (but perhaps not always appropriately) used in quantifying autoradiographs. In this operational mode, two calibrated images may be aligned, and the hybridization signal from one ("non-specific hybridization", for instance) may be digitally subtracted from the other (total hybridization) to produce a digital image of specific signal with grey scales determined from reference to the standard curve. This technique is useful for illustrating specific hybridization (e.g. for publication or other display), which may appear quite different to total signal. In our opinion, however, image subtraction should not be used for the quantification of induced changes in specific hybridization for two reasons. First, it is extremely difficult to perfectly align two images, even using the sophisticated alignment algorithms available in some IASs. Often even serial or near-adjacent sections appear slightly different when placed on slides. Imperfect alignment leads to a reduction in the accuracy of quantification. In addition, the image subtraction procedure is subject to round-off errors, which become more of a problem at the extremes of the calibration scale.

9.4.6 Recording of Image Data

The type of data recorded from film images of ISH is to a large extent dependent on the sampling method used, so sampling should be tailored to the particular application to provide accurate regional measurements or to reveal changes between experimental groups. If cells from a larger area are recruited to express a particular gene following experimental treatments, as is often the case, then it is important to document the area of hybridization signal associated with the regions of interest. In this case, fixed-size sampling tools are inappropriate, and manual or computerized outlining of targets is necessary. Alternatively, or in addition, the average hybridization density may be altered due to changes in the average number of copies of the target mRNA per cell, so under these circumstances density values must be noted to determine the magnitude of any changes seen. We and other researchers often report the density × area product, as this better reflects the total amount of mRNA present and takes into account changes in both determinants (see Fig. 9.2II–III; O'Shea and Gundlach, 1995, 1996; Kowalski *et al.*, 1999; Burazin *et al.*, 2001; Shen *et al.*, 2001). It is up to the individual researcher to ensure that the analysis techniques used enable the reliable detection of a change in cellular mRNA levels in their experimental paradigm if, and only if, such a change exists.

9.4.7 Controls for Use with Film Autoradiography

A number of procedures may be used to assess the specificity of ISH results (Stahl *et al.*, 1993; Wilcox, 1993; and Chapter 1, Section 1.10). When working with a new oligonucleotide or tissue, the researcher should check that hybridization of the radiolabelled probe is displaced by a 10- to 100-fold excess of unlabelled probe (Chapter 1, Section 1.10). A sense probe, equivalent in length and G-C content to the antisense oligonucleotide being used, is also useful in evaluating the authentic nature of the antisense probe hybridization by providing a measure of non-specific hybridization. Multiple oligonucleotide probes, complementary to non-overlapping sections of the target mRNA, are often used to increase hybridization signal strength (Chapter 1, Section 1.8.2 and Fig. 1.11), *and* identical cellular and regional patterns of hybridization with each probe used separately are also a good control for specificity (Wilcox, 1993; and Chapter 1, Section 1.10). For routine analysis using a familiar probe and tissue combination, all these rigid controls are not necessary. In our studies, specific hybridization is recognized as that above the section background signal (which may be due to such factors as chemography, non-specific adhesion of probe or hybridization of 3′-tailed probes to poly(T) sequences in cells). This level of background should be uniformly low in all sections, and should be equivalent to that seen in the presence of an excess of unlabelled probe. Generally, if hybridization has been carried out correctly, regions known *not* to express the transcript being studied have density readings only slightly above film background levels, and signal from these regions is used to define section background, which is subtracted from all readings to give values for specific hybridization. Levels of background can be *relatively* higher in situations where the abundance of the transcript being studied is low (e.g. some transmitter receptors), or where particular individual or combinations of oligonucleotide probes 'stick' more than usual to the tissue (as occasionally observed in our laboratory and others).

9.4.8 Film Exposure Times

Films used in ISH approach saturation at an OD of approximately 1.0. Above this limit, increases in radioactivity produce little or no further increase in labelling density (see Davenport and Hall, 1988; Davenport and Nunez, 1990). For accurate quantification, it is important to control film exposure times to prevent saturation of film over regions of interest. If this practice is not adopted, changes in the abundance of target mRNA may be underestimated or not be detected (e.g. see Chapter 1, Fig. 1.20, where apparent mRNA distribution can change quite a lot, depending on exposure time). Very pale images may be difficult to distinguish from film background, and so multiple exposures may be necessary to quantify areas of widely differing densities, although this is seldom necessary in practice. Regions to be directly compared (i.e. the same region in different animals or multiple regions within animals) should be analysed from a single exposure time, preferably on the same sheet of film if at all possible (see above).

9.5 Quantification of Nuclear Emulsion Images of ISH

9.5.1 Introduction

The study of gene expression at the cellular level, using ISH combined with nuclear emulsion autoradiographic techniques, provides greater anatomical detail than film-based methods. The preparation and analysis of emulsion images of mRNA levels are, however, more technically complex and time-consuming than X-ray film methods. Some researchers regularly analyse emulsion autoradiographs of ISH to provide quantitative information about gene expression, while others employ X-ray films for routine quantification and reserve the use of nuclear emulsions to provide qualitative detail of gene expression at a finer anatomical level. In some experimental situations the broad regional distribution of mRNA is of little relevance, for instance if the mRNA of interest is only expressed in a small number of scattered neurons (e.g. galanin-like peptide (GALP) in the arcuate nucleus; Fig. 9.2IV; Larm and Gundlach, 2000), or it is of low abundance, or if sections are double-labelled in some way (see below and Figs 10.9 and 10.10 in Chapter 10). Under these conditions, emulsion autoradiography of hybridized sections is a far more suitable illustrative and analytical method. Thus, emulsion autoradiography is commonly employed in conjunction with non-isotopic ISH (Chapter 10, Section 10.7), immunohistochemistry, or with conventional histological staining; and quantitative analysis *can* be done under these circumstances. However, it is worth noting that the combination of ISH with other protocols can result in procedural variations in radioactive hybridization levels, due to the nature or length of the additional steps and resultant loss of target mRNA, or induced stripping/quenching of radioactive signal (but see e.g. Cloez-Tayarani and Fillion, 1997, for possible alternative).

9.5.2 Theoretical Aspects of Quantifying Emulsion Images of ISH

ISH combined with emulsion autoradiography allows the detection of the radioactivity concentration in single cells, and thus can provide information about the number of cells expressing the target mRNA and the relative level of expression within individual cells. As is the case for ISH combined with film autoradiography, the number of developed silver grains in the emulsion overlying single cells, within defined limits, is related to the radioactivity concentration (and hence the relative abundance of the target mRNA) in these cells.

While it may be tempting to use emulsion-coated radioactivity standards to produce a standard curve for analysis of emulsion images, and thus calibrate grain densities with known amounts of radioactivity (as for the analysis of film images), this technique is probably best avoided. Radioactivity sources of different shapes and sizes affect emulsions differently (Rogers, 1979), so it cannot be assumed that a homogeneous source of radioactivity (such as a tissue or plastic standard) will produce the same grain density as a smaller source (e.g. a cell) containing the same concentration of radioactivity. For this reason, it is *preferred* to directly determine the density of developed silver grains in the emulsion over areas of the tissue under investigation, and to use these grain-density values as the relative measures of radioactivity content (and hence of target gene expression).

The relationship between the radioactivity concentration of tissue and the density of grains over the tissue is not linear but logarithmic (i.e. the number of β-particles entering the emulsion increases more rapidly than does its grain density), although the first part of this logarithmic curve is approximately linear (Rogers, 1979). Thus, up to a limiting grain density, the relative radioactivity concentration of different samples may be inferred from the density of developed silver grains over the samples, and meaningful comparisons between identically produced samples may be made. The limit of this linearity is related to the size of the undeveloped silver halide crystals in the emulsion, and is reached when ~10% of available crystals have been hit by β-particles (Rogers, 1979). While this theoretical limit may be difficult to determine in practice, the researcher should be aware that, apart from making precise grain counting difficult, significant overlap (or even fusion) of developed grains may severely affect the accuracy of results obtained. Monitoring of emulsion exposure by developing test slides at regular intervals should prevent saturation of the emulsion over the most heavily labelled cells, making analysis of images both easier and more accurate.

In the analysis of emulsion images of ISH, the two questions most commonly being addressed are: how many (or what proportion of) cells express the mRNA of interest (i.e. the number or proportion of radioactively labelled cells), and what is the average relative amount of this mRNA per labelled cell? While it may seem trivial, care must be taken in defining what constitutes a labelled or unlabelled cell. Areas of emulsion over cells that do not produce the mRNA of interest, or in many cases over regions of non-neuronal tissue such as myelin fibre tracts in brain and spinal cord, generally contain low levels of developed silver grains. Since cells which express the target mRNA do so in varying amounts, the frequency distribution of grains per cell (or per unit area) over labelled cells may overlap with the distribution of grain density over unlabelled regions, particularly if the target mRNA is of low abundance. Although complex formulae exist for estimating the proportion of labelled cells (see e.g. Rogers, 1979), simpler methods can also give valid results. One relatively simple formula that can be used to distinguish labelled from unlabelled cells is to set a threshold grain density. This could be equal to or greater than the density over 95% of cells known *not* to express the mRNA in question, or at a level at least 3–5 times higher than background (see e.g. Burton *et al.*, 1992; Bouret *et al.*, 2000). Any cell which has a grain density greater than this limit may then be assumed to be labelled and hence to produce the target mRNA. As is the case with X-ray film, background signal, in the form of (hopefully) low densities of silver grains, should be determined separately and subtracted from total grain density (i.e. specific plus background signal) before the statistical comparison of different samples. For the sake of accuracy, it is preferable that background is kept to a minimum, and attention should be paid to conditions used in the preparation, exposure and development of emulsions to minimize background.

9.5.3 Isotopes and Emulsions

The anatomical resolution that can be achieved using nuclear emulsions is dependent on the isotope and emulsion used, as well as factors such as the section thickness, the length of exposure and the development conditions. Since the radioisotopes most commonly used in ISH (^{35}S, ^{33}P) emit β-particles that have path lengths greater than the emulsion thickness normally achieved (typically 3–4 μm, Rogers, 1979 – using standard dipping procedures, Chapter 1), the even thickness of the emulsion layer is critical for obtaining reproducible results. Standardized procedures for the dipping, drying and development of emulsions are thus essential when these high-energy isotopes are utilized. Emulsion thickness is not critical when ^{3}H-labelled probes are used, since the β-particles from ^{3}H are very unlikely to travel more than 2 μm into the emulsion layer (Rogers, 1979). Differential self-absorption of this isotope by tissue regions of differing density, as well as the long exposure times needed to produce a suitable image, reduce the usefulness of ^{3}H in emulsion ISH. Probes labelled with ^{32}P emit β-particles of sufficiently high energies to cause the production of exposed silver grains at a distance from their source, so the resolution that can be achieved using this isotope is unacceptable for the study of gene expression at the cellular level. While reports exist of the successful nuclear emulsion imaging of hybridization signal within ^{33}P-labelled sections, it is our experience that this isotope performs relatively poorly in this context, particularly with low abundance transcripts. The reason for this is currently unclear (see e.g. Ryan and Gundlach, 1998 for further discussion). In our hands, ^{35}S-labelled sections provide far better results, even with relatively low abundance transcript targets.

Types of Emulsion

The characteristics of the particular nuclear emulsion used, including its grain density, grain size and sensitivity to β-particles of different energies, are also important in determining the resolution that can be achieved. Several photographic supply companies market nuclear emulsions suitable for light microscopic analysis. Emulsions most commonly used in ISH studies are Kodak NTB 2/3, Ilford K5 and Amersham Biosciences LM1/2, which all have undeveloped grain diameters of between 0.20 and 0.26 μm. These emulsions do not require overly long exposure times and can be diluted with water to adjust the emulsion concentration and thickness. We and many others, particularly outside the USA, use Ilford K5 emulsion, diluted 1 : 1 with water/1% glycerol. This is quite economical and gives a uniform emulsion layer of appropriate consistency (see Chapter 1, Protocol 1.9), although Amersham Biosciences LM1/2 and Kodak NTB 2/3 may give slightly quicker results than the Ilford material. Individual emulsions may not be suitable in cases where chemographic artefacts are observed in the presence of some tissue or reagent components (e.g. Trembleau *et al.*, 1993), but to counteract such problems some researchers coat "double-labelled" sections with a thin layer of cellulose (Parlodion) prior to emulsion dipping. This procedure probably does not reduce resolution to a significant degree.

9.5.4 System Requirements for Quantification of Emulsion Images

A variety of techniques are available for the quantitative analysis of emulsion images of ISH, making any attempt to summarize or compare the different methods and the equipment needed to perform them extremely difficult. Grain counting may be performed manually, in which case all that is required is a microscope that allows suitable magnification for the visualization of individual grains. Far more commonly, however, this task is performed with the aid of a computerized IAS. For researchers who already have a microscope and IAS, the decision of which techniques to use may be based on the capabilities of their system, while for those just starting out in the field the number of options may be somewhat bewildering.

The basic equipment needed to quantify ISH at the cellular level comprises a microscope coupled to an IAS. Many of the features of an IAS that are necessary for quantification of emulsion images are identical to those used for analysis of X-ray film images, which are discussed in Sections 9.3 and 9.4.3 (above). The ability to digitally enhance images (to increase the contrast between grains and underlying stained tissue – see below) is an obvious advantage, as is the availability of automated grain-counting procedures. Both these features are common to most IASs. As is the case for the analysis of film images, facilities for the correction of shading errors also aid the accurate quantification of cellular mRNA. Some IASs feature motor stage tracking as an option, allowing the movement of the viewing stage in standard predetermined steps as well as permitting identical areas to be scanned under bright-field (BF) and later under dark-field (DF) conditions (or at different magnifications), without the need to constantly readjust the system when switching between the two viewing modes. The usefulness of a motor stage for some applications must, however, be weighed against its relatively high cost and the sometimes awkward nature of its use.

For the quantification of ISH at the cellular level, attention to constant and standardized analysis conditions is crucial. Typically a light microscope is attached to the video camera of the IAS using a standard video adaptor. Many microscope light sources, for either BF or DF viewing, produce more variation in illumination than the light sources used for analysis of film images. This problem can be overcome to some extent by the use of a line-voltage stabilizer. Most standard light microscopes that have an adjustable light source would be suitable for analysing emulsion autoradiographs of ISH. The ability to view slides under BF and DF conditions and to use a range of objectives to properly visualize silver grains in the emulsion layer are clear advantages. It is important that all components of the optic system (microscope and camera) are clean, since dust or other marks on any of these surfaces can obscure or distort object detail.

9.5.5 Sampling of Emulsion Images

Analysis of emulsion images aims to measure the abundance of developed silver grains over cells of interest, so sampling procedures must distinguish these grains from the underlying tissue. This is achieved using selection criteria that are normally based on the size and darkness of grains (or brightness, depending on the type of illumination used; see below) to highlight grains, which may then be counted or their area measured. In order to distinguish grains from counterstained cells, procedures involving image enhancement or viewing under DF conditions are generally used. The basic sampling procedures using these techniques are described below, and are also outlined in the operation manual of an IAS.

Although the size of individual silver grains in the emulsion layer will vary, the average grain size should be the same for all sections processed and developed identically using one batch of emulsion. For this reason, the area above a correctly chosen threshold density value (i.e. hybridization area) can accurately reflect the number of grains counted manually over a range of grain densities (see e.g. Rogers *et al.*, 1987; Weiss and Chesselet, 1989; Smolen and Beaston-Wimmer, 1990). It also has the advantage of avoiding the uncertainty in discerning individual grains when some overlap or fusion occurs. In order to use measurement of the area occupied by grains, the researcher should determine the limits between which a linear relationship between grain number and grain area exists. These limits will depend on factors such as the magnification used, resolving power of the IAS, and the types of emulsion used. Manual grain counts and computerized area measurements (pixels or calibrated area) from cells having a wide range of grain densities should be used to construct a calibration curve relating grain number to grain area. Linear regression analysis can be used to determine a line of best fit, the slope of which is then used as a conversion factor to relate grain density to area above threshold for all cells from a particular experiment. Under a range of analysis conditions, a linear relationship between grain density *and* area has been demonstrated for 10–150 grains per cell (Rogers *et al.*, 1987; Weiss and Chesselet, 1989; Smolen and Beaston-Wimmer, 1990).

Emulsion Autoradiograph Analysis under Bright-field Illumination

Under BF conditions, developed silver grains appear as black specks above the stained tissue. Light counterstaining of cells so that cell detail is only just visible aids in reliably distinguishing grains from the underlying tissue, which is critical to the analysis under these conditions. Image enhancement operations, which are a feature of most IASs, also greatly increase the contrast between cells and grains. These procedures modify images by changing the intensity value of each pixel, basing the new value on those of the surrounding pixels. In the case of grain counting, small dark objects (grains) are highlighted relative to larger pale objects (cells) using target accentuation or sharpening filters (Fig. 9.3). More sophisticated image enhancement, discriminating targets by their size and shape, may be used to separate contiguous targets (e.g. overlaid silver grains) before counting. Image-processing operations do, by their nature, alter image data, so the researcher should ensure that any processing operations carried out do not bias results in a way that will alter the outcome. An example of the analysis of emulsion images under BF viewing conditions is given in Protocol 9.2.

Emulsion Autoradiograph Analysis under Dark-field Illumination

The use of DF illumination to quantify emulsion images of ISH provides greater contrast between developed silver grains and stained tissue, and may avoid the need to use image-processing routines to distinguish grains. Under these conditions, bright silver grains are visible over a dark background, with cells faintly visible due to their counterstaining. Lewis *et al.* (1989) report that more reliable quantification could be achieved with DF than BF illumination, using the same magnification. One disadvantage of DF viewing is that slides and sections must be thoroughly clean to avoid artefacts caused by dust and other contaminants.

Analysis of emulsions under DF illumination is carried out in a manner

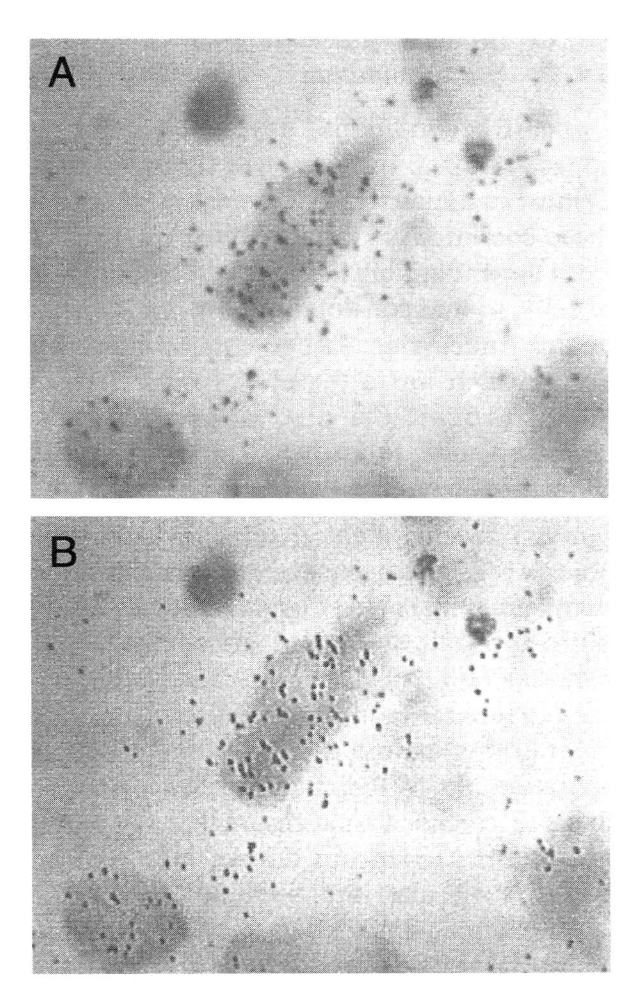

FIG. 9.3. Image enhancement to increase the contrast between cells and silver grains. Emulsion autoradiographic images shown illustrate lateral hypothalamic cells expressing somatostatin mRNA. (A) Original image using 100× objective and oil immersion. Note low contrast between silver grains and stained cells, as well as some overlap of grains. (B) Same field as in A after image enhancement algorithm to highlight small dark objects. Grains are now clearly distinguishable from background.

almost identical to that described in Protocol 9.2 for BF, with the exception that bright rather than dark targets are selected by threshold determination. As is the case for BF viewing, a highly significant correlation between manually counted grains and computer-generated pixel counts can be achieved with DF illumination (Rogers *et al.*, 1987).

Protocol 9.2 Analysis of Emulsion Autoradiographs under BF Conditions

1. Select magnification and adjust condenser. Individual grains can be distinguished, and the relative density of grains over different regions can be compared using 40× to 100× objective lenses (with oil immersion for 100×). Higher magnifications allow greater numerical accuracy by increasing the pixel/grain ratio, but slow down analysis by necessitating frequent movement of the viewing field. Check that the viewing field is of appropriate size for the tissue under investigation, and focus the microscope on the emulsion layer. Condenser aperture and height should be appropriate for the objective being used (see microscope manual).
2. Shading error correction to remove non-uniform illumination conditions over the field of view should be carried out without a slide present, after the microscope has been focused.
3. Find an appropriate field of view to quantify. The researcher should decide (preferably before the start of sampling) which (and how many) cells to quantify in order to gain meaningful and representative data. The abundance and distribution of the target mRNA species, as well as the number of sections or animals available for study, will obviously influence the number and type of cells from different regions which can be sampled. It is important to have clear goals in mind before sampling in order to ensure sufficient and appropriate data are obtained. Since cells in each section will be transected at different levels and hence will have widely differing cross-sectional areas, it may be advisable to adopt some standard inclusion and exclusion criteria, for example only sampling from cells which have the nucleus in view.
4. Check focus each time the field of view is adjusted, to ensure that most grains are sharp. At high magnification, some grains will appear out of focus, but this is unavoidable.
5. Digitize image and perform image enhancement procedures (if necessary) to increase contrast (Fig. 9.3).
6. Set OD threshold for analysis. After image enhancement, grains should be easily distinguishable from the underlying tissue. Threshold is adjusted so that grains are selected while underlying tissue is not. Adjusting the threshold for each cell or field of view, rather than using the same threshold value for all measurements, compensates for variations in lighting and in the staining density of underlying tissue, and so gives more reproducible results (e.g. Lewis *et al.*, 1989).
7. Define the region(s) to be sampled. The drawing and geometric tools described in Section 9.4.5 may be used to select areas of the field to be analysed. With ^{35}S and isotopes of similar energy, some scatter of silver grains around the actual cell is observed, so a standard method of defining target areas must be devised. One example would be to draw a border at a set distance outside the cell boundary for each cell to be analysed.
8. Sample from defined area. Automated grain- or pixel-counting procedures give a value for the number of grains or pixels above threshold.

Since the size of cells may change due to some experimental treatments, grain or pixel *density*, rather than just number, is a more reliable measure of mRNA abundance, and requires the recording of the area sampled as well as the grain or pixel count for the area. Thus, for each sampling region, the number of grains per unit area or the proportion of the total sampling area above threshold (area above threshold/total area) may be recorded.

9. Sample next area (repeat steps 1–8).

9.5.6 Controls for Use with Emulsion Autoradiography

As well as employing control strategies to assess the specificity of hybridization (see Section 9.4.7, and Chapter 1, Section 1.10), the researcher should also follow a number of procedures to ensure the validity of results obtained from emulsion autoradiography (Rogers, 1979). Silver grains may be produced in the emulsion by a number of sources other than the radioactively labelled probe that is hybridized to tissue sections. Thus false-positive results should be controlled for by the emulsion dipping and development of non-radioactive (non-hybridized) sections along with every batch of experimental slides. In addition, silver grains may be lost from areas of specific hybridization (i.e. false-negative results). To control for this occurrence, a hybridized slide should be exposed to light at the time of dipping and then exposed and processed with experimental slides. Any loss of emulsion response over the hybridized tissue indicates (most commonly) chemography or fading of the latent image during exposure (Rogers, 1979), factors which could seriously affect the accuracy of quantifying other sections.

9.6 Quantification of Non-isotopic ISH

A further innovation in the study of gene expression is the use of non-isotopic ISH methods (see Chapters 10–12). These techniques offer improved cellular (and regional) resolution and faster results when compared with conventional radioactive methods, and dispense with the need for safety procedures that must be observed when radioactivity is used. The use of non-isotopic techniques in *quantitative* ISH has not yet become widespread (e.g. Larsson *et al.*, 1991), although their popularity is growing, as simpler and higher sensitivity methods are devised (see e.g. Bouret *et al.*, 2000; Landry *et al.*, 2000).

Non-isotopic ISH most commonly involves the labelling of the probe (oligonucleotide, cDNA or cRNA) with a marker, such as digoxigenin (Chapters 11 and 12) or biotin, by the incorporation of a modified nucleotide. An alternative method involves the use of an enzyme (alkaline phosphatase or horseradish peroxidase) that is conjugated to oligonucleotide probes by a 'linker arm' molecule during synthesis (Jablonski *et al.*, 1986; Emson, 1993; Chapter 10). The reporter molecules are then utilized to produce a coloured or fluorescent reaction product, the abundance of which may be analysed quantitatively (Chapter 10; see e.g. Larsson *et al.*, 1991; Emson, 1993; Landry *et al.*, 2000). Recently, additional amplification detection procedures, such as the biotin-tyramide signal amplification (TSA; NEN) method, have become commercially available and offer the user a potential (significant) increase in sensitivity of detection of hybridization signal with biotin-labelled probes. At this stage, however, it is not clear whether their use will become as widespread as the earlier alkaline phosphatase-based method, or what the relative sensitivities of these methods are in various experimental situations and when using different probe types.

In using non-isotopic methods in ISH, the same types of procedures as those used in conventional ISH (Table 9.1) should be observed to minimize the variability of results. Since each step involved in the processing of sections and the amplification of signal introduces more chance of variability, the use of simpler procedures, such as those utilizing AP-labelled oligonucleotide probes (Augood *et al.*, 1992; Chapter 10) *may* give more reproducible results for quantitative studies, without a significant loss of sensitivity. However, this approach is more expensive and less flexible than methods involving 3′-labelling of probes with modified nucleotides, using the terminal transferase enzyme.

Sections processed for non-isotopic ISH may be used to gain information about the number (or proportion) of cells expressing the target mRNA as well as the average relative level of labelling of these cells. Briefly, a microscope set to an appropriate magnification and coupled to a computerized IAS may be used to analyse labelled cells in a manner similar to that used for the study of emulsion images of ISH (Section 9.5). If relative cellular OD measurements are to be recorded, colour development should be carefully monitored to ensure that overdevelopment (analogous to film or emulsion saturation in isotopic ISH) does not occur (Larsson *et al.*, 1991). Since no external standards are used in these procedures, it may not be appropriate to compare data across experiments. Sections from individual experiments may be compared with each other, and changes in the number of positive cells or the degree of labelling observed following experimental treatments (e.g. Chapter 10; Augood *et al.*, 1992; Landry *et al.*, 2000).

Non-isotopic ISH may also be used to study gene expression at the regional level, using analysis techniques akin to those used for film images

(Section 9.4). Once again, the absence of external standards limits the validity of comparisons made across experiments using this method (but see comments regarding probing with the identical probe batch in Chapter 11, Section 11.1.2).

9.7 Statistical Analysis and Reporting of Results

Appropriate statistical analysis of results from ISH and the ability to describe these findings accurately are important and often overlooked aspects of the study of gene expression using this technique. As more researchers (and journal referees) gain expertise in ISH *and* statistical methods, the "one-test-fits-all" mentality in molecular biosciences is being replaced by the use of appropriate statistical tests. A detailed description of statistical methods is beyond the scope of this article, and the reader is advised to refer to any of the many appropriate texts on biostatistics and analysis (e.g. Sokal and Rohlf, 1981; Armitage and Berry, 1987).

9.7.1 Some Notes on Statistics in the Quantification of ISH

There is seldom consensus between researchers on the exact methods of statistical analysis that are appropriate for their experimental results, and this state reflects both the broad array of possibilities and the absence (in most cases) of a single "best" solution. Many different statistical software packages are available to the researcher, and the capabilities of the programs available often dictate which statistical methods are used, or alternatively some programs actually recommend to the user which type of analysis is best.

The use of ISH in experiments where comparisons are made between data from animals in a number of treatment groups (including control or sham-treated groups) is a common application, and some guidelines for the analysis of these experiments are presented here. First, films or emulsions should *preferably* be sampled by an observer "blind" to the treatment groups, to prevent operator bias from influencing image analysis. Randomized coding of animals across groups, with the code being revealed following sampling, is a simple way to achieve this objective, although practically this is not always possible. Following image analysis, data (average or mean ± SEM for the parameters being studied) from each sampling region from all animals in a particular treatment group may be pooled to give group statistics (mean ± SEM). Group data may then be compared in many ways. If a number of groups are being compared with a control group and/or with each other, multiple comparison procedures, which minimize the risk of making false inferences from multiple comparisons, should be employed (see e.g. Ludbrook, 1991, 1998 for review). Since the variation between samples of biological material is often greater than that predicted by the normal distribution, non-parametric statistical procedures might be appropriate in many situations (Siegel, 1956).

9.7.2 Reporting of ISH Methods and Results

There is no standard method for the reporting of ISH experiments, and large variations in the amount of information presented are seen in the literature. Since readers of reports dealing with ISH should be able to replicate the findings presented, more rather than less detail in the description of methods is an advantage. To this end, researchers should accurately describe, or refer to accurate descriptions of, all the experimental and analytical methods employed. This attention to detail should also apply to control procedures. If the researcher has gone to the trouble of ensuring the specificity of the results presented, these details should be provided. Since, as previously explained, so many options are available for image and statistical analysis of results, the procedures adopted for these important tasks also need to be detailed.

In reporting the results of ISH experiments, an accurate description of the findings seen, along with appropriate examples of the raw data obtained (i.e. conventional photographs or digital images of film or emulsions), are invaluable in conveying these results to the reader. Explanations (including statistics) of any experimental changes seen should also be accompanied by illustrations of these effects, in addition to appropriate graphs or tables.

9.7.3 Relative versus Absolute Measurements of MRNA Abundance

The methods described in this chapter for ISH analysis are essentially only useful for the *relative* quantification of regional or cellular mRNA. It is possible, if the specific activity of the probe and the efficiency of autoradiography are known, to calculate the number of hybrids/cell from film OD or emulsion grain density (see Lewis *et al.*, 1989; Young, 1990). Relating this value to the absolute mRNA concentration is, however, extremely difficult and relies on a number of assumptions. In the case of ISH, the loss of mRNA during preparation of samples, the efficiency of hybridization and the fact that transected cells or neurons are used in hybridization, are all factors that must cast doubt on the value of any absolute measurements. In addition, the response of X-ray film or nuclear emulsion to radioactivity is often unpredictable due to chemography, latent image fading and, if tritium is used, differential quenching (Rogers, 1979), further impeding attempts to obtain absolute measurements. In contrast, the relative quantification of ISH, using methods like those presented here, has been extensively used by many research groups to provide meaningful and accurate comparisons of mRNA (gene expression) levels in many different tissues, following a range of experimental treatments and during the course of brain development. Often the validity of these results is confirmed by independent and/or complementary methods.

9.8 Reflections and Future Directions

Since the publication of the first edition of this book in 1994, we have observed a consistent or increasing number of published articles in neuroscience and other disciplines utilizing the ISH technique. Many of these papers from laboratories with expertise in molecular biology rely on PCR- or plasmid-derived cRNA probes, particularly when using non-radioactive ISH. Those from "non-molecular" laboratories often use more convenient oligonucleotide probes. Irrespectively, over the years, ISH, both qualitative and semiquantitative, has become a method used in conjunction with a number of other complementary techniques, with fewer papers publishing *only* ISH data. Thus, the trend appears to be that studies reporting the identification/cloning, regulation and/or function of a novel gene/protein routinely examine the distribution of gene expression across a range of tissues by Northern blot or RT-PCR analysis and expression *within* tissues, using ISH. However, the importance of mapping cellular and regional gene expression and its regulation is also being increasingly realized with so many new genes being rapidly identified. So much so that there is even a journal, *Gene Expression Patterns* (Elsevier), and at least one database devoted to the topic. We believe that in the future, quantitative ISH will continue to play a valuable role.

9.9 Conclusion

The technique of ISH is a powerful tool in the study of gene expression, and is in increasingly widespread use in the biological sciences and particularly in the field of neuroscience. The desire to measure the extent of apparent differences between experimental samples has led to the development of sophisticated, but easy-to-use, computer analysis procedures, some of which have been discussed in this chapter. Alterations in gene expression (as reflected by mRNA levels) involving an approximate halving or doubling (or more) of control mRNA levels have regularly been reported by many research groups in a wide variety of experimental models. Moreover, despite the inherent variability of the experimental techniques used in ISH, changes as small as 20–30% can be documented using properly applied ISH in appropriate-sized groups, in combination with accurate computerized image analysis. With anticipated advances in these technologies, the reproducibility, sensitivity and affordability of these techniques will no doubt improve even further in the future.

Acknowledgements

Research in the author's (ALG) laboratory over the last decade has been supported by grants from the National Health and Medical Research Council of Australia and the Austin Hospital Medical Research Foundation. Substantial equipment grants to assist in the purchase of a new image analysis system have also been received from the Sylvia and Charles Viertel Charitable Trust, the Clive and Vera Ramaciotti Foundations, the Ian Potter Foundation, the Jack Brockhoff Foundation and the University of Melbourne. I would like to take this opportunity to acknowledge the valued contributions of my postgraduate students and postdoctoral scholars to various studies of gene regulation in the nervous system, completed over this time.

References

Armitage, P. and Berry, G. (1987). "Statistical Methods in Medical Research". Blackwell Scientific Publications, Oxford.

Augood, S. J., Faull, R. L., and Emson, P.C. (1992). *Eur. J. Neurosci.* **4,** 102–112.

Bhat, R. V., Cole, A. J., and Baraban, J. M. (1992). *J. Pharmacol. Exp. Ther.* **263,** 343–349.

Bouret, S., Prevot, V., Croix, D., Howard, A., Habert-Ortoli, E., Jegou, S., Vaudry, H., Beauvillain, J. C., and Mitchell, V. (2000). *Endocrinology* **141,** 1780–1794.

Burazin, T. C. D., Larm, J. A., and Gundlach, A. L. (2001). *J. Neuroendocrinol.* **13,** 358–370.

Burton, K. A., Kabigting, E. B., Clifton, D. K., and Steiner, R. A. (1992). *Endocrinology* **131,** 958–963.

Cloez-Tayarani, I. and Fillion, G. (1997). *Brain Res. Protoc.* **1,** 195–202.

Dagerlind, A., Friberg, K., Bean, A. J., and Hökfelt, T. (1992). *Histochemistry* **98,** 39–49.

Davenport, A. P. and Hall, M.D. (1988). *J. Neurosci. Methods* **25,** 75–82.

Davenport, A. P. and Nunez, D. J. (1990). *In* "*In Situ* Hybridization: Principles and Practice" (J.M. Polak and J. O'D. McGee, Eds), pp. 95–111. Oxford University Press, Oxford.

Eakin, T. J., Baskin, D. G., Breininger, J. F., and Stahl, W. L. (1994). *J. Histochem. Cytochem.* **42,** 1295–1298.

Eilbert, J. L., Gallistel, C. R., and McEachron, D.L. (1990). *Comp. Med. Imag. Graph.* **14,** 331–339.

Emson, P. C. (1993). *Trends Neurosci.* **16,** 9–16.

Harlan, R. E., Shivers, B. D., Romano, G. J., Howells, R. D., and Pfaff, D. W. (1987). *J. Comp. Neurol.* **258,** 159–184.

IHGSC (International Human Genome Sequencing Consortium) (2001). *Nature* **409,** 860–921.

Jablonski, E., Moomaw, E. W., Tullis, R. H., and Ruth, J. L. (1986). *Nucleic Acids Res.* **14,** 6115–6128.

Kowalski, T. J., Houpt, T. A., Jahng, J., Okada, N., Liu, S. M., Chua, S. C. J., and Smith, G. P. (1999). *Physiol. Behav.* **67**, 521–525.

Kuhar, M. J., De Souza, E. B., and Unnerstall, J. R. (1986). *Annu. Rev. Neurosci.* **9,** 27–59.

Kuhar, M. J., Lloyd, D. G., Appel, N., and Loats, H. L. (1991). *J. Chem. Neuroanat.* **4,** 319–327.

Landry, M., Holmberg, K., Zhang, X., and Hökfelt, T. (2000). *Exp. Neurol.* **162,** 361–384.

Larm, J. A. and Gundlach, A.L. (2000). *Neuroendocrinology* **72,** 67–71.

Larsson, L. I., Traasdahl, B., and Hougaard, D. M. (1991). *Histochemistry* **95,** 209–215.

Lewis, M. E., Rogers, W. T., Krause, R. G. I., and Schwaber, J. S. (1989). *Methods Enzymol.* **168,** 808–821.

Ludbrook, J. (1991). *Clin. Exp. Pharmacol. Physiol.* **18,** 379–392.

Ludbrook, J. (1998). *Clin. Exp. Pharmacol. Physiol.* **25,** 1032–1037.

McCabe, J. T., Kao, T.-C., and Volkov, M. L. (1993). *Microsc. Res. Tech.* **25,** 61–67.

Miller, J. A. (1991). *Neurosci. Lett.* **121,** 211–214.

Mize, R. R., Thouron, C., Lucas, L., and Harlan, R. (1994). *Neuroimage* **1,** 163–172.

O'Shea, R. D. and Gundlach, A. L. (1994). *In* "*In Situ* Hybridization Protocols for the Brain" (W. Wisden and B. J. Morris, Eds), pp. 57–78. Academic Press, London.

O'Shea, R. D. and Gundlach, A. L. (1995). *Peptides* **16,** 1117–1125.

O'Shea, R. D. and Gundlach, A. L. (1996). *J. Neuroendocrinol.* **8,** 417–425.

Paxinos, G. and Watson, C. (1986). "The Rat Brain in Stereotaxic Coordinates". Academic Press, Sydney.

Ramm, P. (1990). *Comp. Med. Imag. Graph.* **14,** 287–306.

Rogers, A. W. (1979). "Techniques of Autoradiography". Elsevier, New York.

Rogers, W. T., Schwaber, J. S., and Lewis, M. E. (1987). *Neurosci. Lett.* **82,** 315–320.

Ryan, M. C. and Gundlach, A. L. (1998). *Dev. Brain Res.* **105,** 251–268.

Shen, J., Larm, J. A., and Gundlach, A. L. (2001). *Neuroendocrinology* **73,** 2–11.

Shen, P.-J. and Gundlach, A. L. (1998). *Exp. Neurol.* **154,** 612–627.

Shen, P.-J. and Gundlach, A. L. (1999). *Exp. Neurol.* **160,** 317–332.

Siegel, S. (1956). "Non-parametric Statistics". McGraw-Hill, New York.

Smolen, A. J. and Beaston-Wimmer, P. (1990). *In* "*In Situ* Hybridization Histochemistry" (M.-F. Chesselet, Ed), pp. 175–188. CRC Press, Boca Raton.

Sokal, R. R. and Rohlf, F. J. (1981). "Biometry". W. H. Freeman, New York.

Stahl, W. L., Eakin, T. J., and Baskin, D. G. (1993). *J. Histochem. Cytochem.* **41,** 1735–1740.

Trembleau, A., Roche, D., and Calas, A. (1993). *J. Histochem. Cytochem.* **41,** 489–498.

Venter, J. C., Adams, M. D., Myers, E. W. *et al.* (2001). *Science* **291,** 1304–1351.

Weiss, L. T. and Chesselet, M.-F. (1989). *Mol. Brain Res.* **5,** 121–130.
Wilcox, J. N. (1993). *J. Histochem. Cytochem.* **41,** 1725–1733.
Young W.S., III (1990). *In* "Handbook of Chemical Neuroanatomy, Vol. 8 Analysis of Neuronal Microcircuits and Synaptic Interactions" (A. Björklund, T. Hökfelt, F. G. Wouterlood and A. N. van den Pol, Eds), pp. 481–512. Elsevier, Amsterdam.
Zhang, B., Glasgow, E., Murase, T., Verbalis, J. G., and Gainer, H. (2001). *J. Neuroendocrinol.* **13,** 29–36.

PART II

NON-RADIOACTIVE *IN SITU* HYBRIDIZATION

CHAPTER 10

NON-RADIOACTIVE *IN SITU* HYBRIDIZATION USING ALKALINE PHOSPHATASE-LABELLED OLIGONUCLEOTIDES

S. J. Augood,* E. M. McGowan,** B. R Finsen,† B. Heppelmann‡ and P. C. Emson§

*Neurolology Research, Massachusetts General Hospital, Charlestown, MA 02129, USA
**Birdsall Building, Mayo Clinic, Jacksonville, Florida, USA
†Anatomy and Neurobiology, University of Southern Denmark – Odense University, Winslowparken 21, DK-5000 Odense C, Denmark
‡Physiologisches Institut, Universität Würzburg, Röntgenring 9, D-97070 Würzburg, Germany
§Laboratory of Molecular Neuroscience, The Babraham Institute, Cambridge CB2 4AT, UK

10.1 Introduction to Alkaline Phosphatase-labelled Oligonucleotides

In the search to make classical radioactive *in situ* hybridization (ISH) methods using 3′-tailed synthetic oligonucleotides less complex and more accessible to researchers without radiochemical facilities, various non-radioactive methods have been developed. In most cases, the synthetic oligonucleotide is tagged with a reporter molecule, such as biotin or mercury, at the 3′ or 5′ end (Langer *et al.*, 1981; Agrawal *et al.*, 1986; Hopman *et*

ISSN/02 $00.00

al., 1986; Emson, 1993), and then the sites of hybridization are visualized using a suitable detection system, for example, a streptavidin antibody complex. Although many of these "indirect" *in situ* methods are sensitive and give excellent results on membrane blots and tissue sections, they may be more suited to qualitative rather than quantitative studies, as the detection procedure involves several amplification steps.

In our experience, for both qualitative and semiquantitative studies, alkaline phosphatase (AP) is the reporter of choice for labelling oligonucleotides (e.g. Figs 10.1, 10.7 and 10.8). Although these probes offer many advantages over their non-radioactive adversaries, and indeed radiolabelled oligonucleotides, they do have their limitations. AP probes (i) give excellent cellular resolution of hybridization sites on tissue sections (Kiyama and Emson, 1990; Kiyama *et al.*, 1990a; Augood *et al.*, 1991a,b, 1992; Augood and Emson, 1992; West *et al.*, 1996; Augood *et al.*, 1997; Hougaard *et al.*, 1997; Pedersen *et al.*, 1998; Andreassen *et al.*, 1999; Ohnuma *et al.*, 1999; Gregersen *et al.*, 2001; Lambertsen *et al.*, 2001) and neuronal cultures (Finsen *et al.*, 1992; Østergaard *et al.*, 1995) (a particulate signal is detected in the cell cytoplasm, the definition of the signal being dependent upon the thickness of the tissue section, the thinner the section the more cellular detail is visible), mRNA can be detected in cell processes (Fig. 10.7 D, E), (ii) may be purchased commercially and stored at 4°C for at least 12 months without loss of enzyme activity, (iii) are easy to use (no additional detection kit/ antibody complex is required to detect sites of probe hybridization), (iv) are fast (the AP hybridization signal is detected within 12–48 h after washing of the sections), (v) may be used for semiquantitative analysis as no amplification steps are involved (the intensity of the AP hybridization signal is directly proportional to the amount of probe hybridized (therefore the amount of mRNA present)), (vi) may be combined with radioactive oligonucleotides (^{35}S and possibly ^{33}P) and antibodies for co-expression and co-localization studies, and (vii), in our hands at least, are more sensitive than other non-radioactive counterparts.

The disadvantages are less numerous, yet just as important: (i) these oligonucleotides are not as sensitive as ^{35}S-labelled oligonucleotides when detecting rare transcripts, (ii) they will not work to give the optimal hybridization signal without modification of the standard hybridization conditions given in Chapter 1, (iii) they are relatively expensive to purchase (the purity of the labelled probe varies considerably and is dependent on the supplier), and (iv) although several "user-friendly" labelling kits are available, in our experience, the end product often requires further purification by e.g. fast protein liquid chromatography (FPLC).

10.1.1 Synthesis and Labelling of AP-oligonucleotides

In 1986 Jablonski and colleagues reported a method for covalently cross-linking calf intestinal AP to short (21–26mers) synthetic oligonucleotides. This was one of the first "direct" non-radioactive methods to be reported. A simplified overview of their method of labelling with AP is given below; some modifications have been incorporated. A modified thymine base terminating in a reactive primary amine is incorporated directly into the automated oligonucleotide synthesis; the modified oligonucleotide is then reacted with the homobifunctional reagent disuccinimidyl suberate (pH > 7), and the reaction allowed to proceed for 5 min in the dark; the reaction mixture is then applied to a Sephadex G-25 column, eluted with water, immediately frozen and lyophilized. The activated linker-arm oligonucleotide is then rehydrated with a 2-fold excess of AP and the conjugation reaction is left to proceed for 16 h at room temperature. Finally protein products are separated from non-protein components by gel filtration, while the pure AP-oligonucleotide conjugate is separated from free AP by FPLC. The purified end product is then concentrated using a Minicon Macrosolute Concentrator (Amicon). This chemical procedure is discussed in detail elsewhere (Kadowaki *et al.*, 1993).

10.1.2 Storage of AP-oligonucleotides

AP-oligonucleotides should not be frozen and should be stored at 4°C, as recommended by the manufacturers. To preserve the activity of the enzyme, we store our AP-oligonucleotides in buffer containing 30 mM Tris, 3.0 M NaCl, 1.0 mM $MgCl_2$, 0.1 mM $ZnCl_2$ and 0.05% sodium azide, pH 7.6 at 4°C. Once these oligonucleotides have been diluted in hybridization buffer, they should be used immediately. They should *never* be stored for long periods in the presence of formamide, for example diluted in hybridization buffer, because formamide (> 36 h) significantly reduces the activity of the enzyme and therefore reduces the intensity of the final hybridization signal.

10.2 Preparation of Fresh Frozen CNS Tissue Sections

Once the brain has been removed from the skull, and frozen as described in Chapter 1 (Section 1.5), we have found it best to thaw-mount cryostat sections (10–15 μm) on to gelatin–chrom alum-coated slides. Alternative

PROTOCOL 10.1 PREPARATION OF GELATIN–CHROM ALUM-COATED SLIDES.

1. Wearing gloves, load microscope slides[a] into glass staining racks.[b]
2. Rinse slides in acetone (5 min) and air-dry.
3. Warm 500 ml of DEPC-treated water to 50°C then add 5 g of gelatin powder and 0.25 g of chromic potassium sulphate. Stir until dissolved.
4. Filter solution and allow to cool to 37°C; dip slides in glass slide racks into the warm gelatin solution for 3 min, remove and drain off excess.
5. Leave slides to dry in an oven for several hours.
6. Coated slides may be stored for several weeks in partitioned slide boxes.

[a] For convenience we use twin-frosted precleaned slides from Solmedia.
[b] It is advisable to keep a set of glass slide racks which are used exclusively for coating as the gelatin hardens on to the racks and a substantial deposit can build up. Gelatin–chrom alum solution may be reused. Filter the solution, autoclave and store at 4°C. Before subbing the slides, warm the gelatin solution to 37–42°C (we routinely do this in a microwave).

adhesion substrates may be used, such as poly-L-lysine or Vectabond (Vector Labs); however, silating agents tend to inhibit AP enzyme activity and can result in a reduced AP hybridization signal. The recipe for preparing gelatin-coated slides is given in Protocol 10.1.

10.2.1 PREPARATION OF PARAFFIN-EMBEDDED CNS TISSUE SECTIONS

For qualitative studies of gene expression, paraffin-embedded semi-thin (5–7 µm) sections offer excellent cellular resolution of the AP hybridization signal combined with preserved tissue morphology (see Fig. 10.1). Furthermore, after development of the coloured AP reaction product, sections may be processed for immunocytochemistry allowing antigens and mRNA to be simultaneously visualized in the same tissue section (see Section 10.8; Emson *et al.*, 1993; Heppelman *et al.*, 1994). In our experience it is preferable to carry out the ISH first and the immunocytochemistry subsequently; however, this ordering depends on the stability of the antigen; for the localization of oestrogen receptor immunoreactivity in the cell nucleus, for example, it may be better to carry out the immunocytochemistry procedure first (using sterile conditions and diethylpyrocarbonate (DEPC)-treated buffers) and the AP ISH second (A. Herbison, personal communication). The ordering of the two steps should be determined for each antibody. We have had most success performing the immunocytochemistry second.

Paraffin-embedded blocks are prepared using a routine method. For CNS tissue, rodents are first anaesthetized with Sagattal, perfused transcardially

PROTOCOL 10.2 PARAFFIN EMBEDDING OF TISSUE BLOCKS (ROUTINE PROCESSING BY MACHINE).

1. Fix tissue in 4% paraformaldehyde for up to 24 h (largely dependent on size of tissue).
2. Dehydrate through ethanol: 50% (12 h), 70% (12 h), 90% (3 h), 2 × 100% (1.5 h each), fresh 100% (2 h), then delipidate in chloroform: I, 3 h; II, 8 h.[a]
3. Impregnate tissue with paraffin wax (60°C): use three changes of wax, 1 h each (last change in a vacuum bath) before finally embedding in fresh wax (60°C) in a cast.
4. Allow wax to set, remove wax block from plastic moulding and trim excess paraffin. Before cutting, allow block to stand on ice (cut face down).
5. Cut ribbons of 5–7 μm sections, float out on to the surface of a warm water bath (to allow sections to unravel), when sections completely flat, mount on to coated slides[b] and dry. Sections may then be stored at room temperature for several months.

Rapid processing for small tissue (e.g. dorsal root ganglia)

1. Fix tissue in 4% paraformaldehyde[c] and dehydrate: 50% (1 h), 70% (1 h), 90% (1 h), 100% twice 45 min each.
2. Delipidate tissue with benzene/toluene twice 15–45 min per change.
3. Impregnate with warm paraffin wax 2 × 1 h, then embed in fresh wax as above.

[a] Tissue may stay in chloroform for longer than 8 h.
[b] Gelatin–chrom alum-coated slides are not recommended here as the sections are already fixed, so tissue proteins will not cross-link with the gelatin and hold the sections in place; consequently sections will float off. Alternative substrates that may be used for paraffin sections include Vectabond (Vector Labs) or 3-aminopropyltriethoxysilane (APES; Sigma; A-3648).
[c] If tissue is fixed in Bouin's fluid, then miss out 50% ethanol and go straight to 70% ethanol.

with freshly prepared ice-cold heparinized (1%, v/v) DEPC-treated saline followed by 4% neutral buffered paraformaldehyde in 0.1 M phosphate-buffered saline (PBS); tissue blocks are then postfixed overnight. Submersion of fixed tissue may also be used, for example, post-mortem human blocks, but the morphology is, as expected, not as well preserved. Blocks are then processed as standard (see Protocol 10.2).

There are several advantages to be gained from using paraffin-embedded tissue for ISH studies: (i) the morphology of the sections is excellent and allows clear observation of the cellular localization of the AP hybridization signal, usually seen as a particulate "hot spot" within the cell cytoplasm of thin (5–7 μm) sections; (ii) it allows access to a vast archive of human

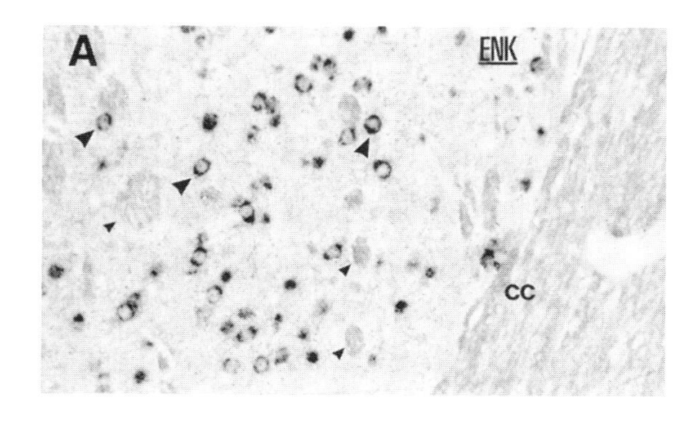

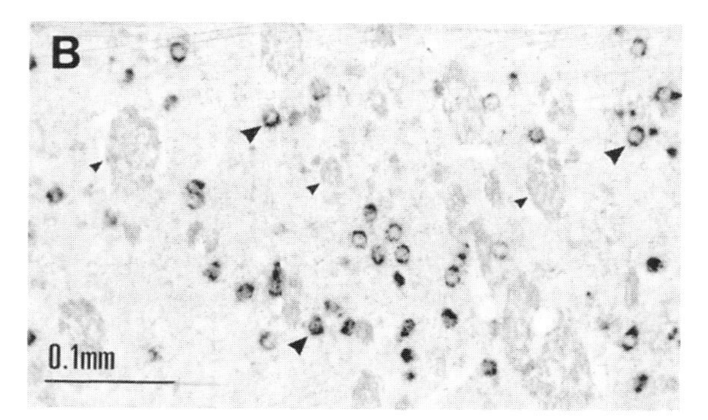

FIG. 10.1. Semi-thin (7 μm) sections of paraffin-embedded rat striatum hybridized with AP-labelled oligonucleotide complementary to a portion of the rat pre-proenkephalin A (ENK) mRNA. The AP reaction product is concentrated in the cell cytoplasm whereas the cell nuclei are relatively devoid of signal, illustrated by large arrowheads in both (A) and (B); the smaller arrowheads indicate fibre bundles in the striatum. cc, corpus callosum.

post-mortem tissue collected and processed for routine histological studies; and (iii) tissue sections may be cut, mounted on to RNase free coated slides and stored at room temperature for several months (if not longer) without any deterioration in hybridization signal.

10.2.2 PREPARATION OF NEURONAL CULTURES

When applying AP ISH to neuronal cultures (see Fig. 10.2 for example), it is important to ensure that (i) the cultures are grown on a support medium which may be used directly in the hybridization process, (ii) that this support medium is pretreated with a substrate that will ensure that the cultures

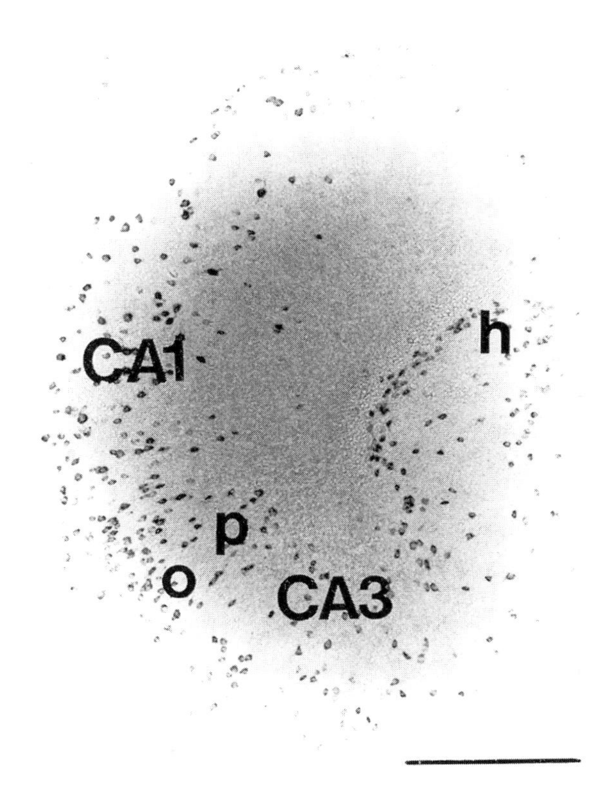

FIG. 10.2. Somatostatin (SRIF) gene expression in 5-week-old whole hippocampal explant culture. Note the high numbers of organotypically distributed neurons throughout the hippocampal fields. The number and distribution of neurons detected by the AP-SRIF probe is, in essence, identical with that revealed by immunocytochemistry using an anti-SRIF antibody (Finsen *et al.,* 1992). h, dentate hilus; o, stratum oriens; p, stratum pyramidale hippocampal fields ca1 and ca3. Scale bar 400 μm.

will not float off during the relatively high-temperature (55°C) post-hybridization washes, and (iii) that this substrate is relatively inert and will not adversely affect the developing cultures. To this end we grow both dispersed primary cultures and CNS slice explants on poly-L-lysine-coated glass coverslips (Chapter 1, Protocol 1.1). Routinely, explants are from P1–P7 Wistar rat pups; tissue slices are cut (350 μm) using a McIlwain tissue chopper and then placed individually on coverslips in a drop of chicken plasma, coagulated by thrombin and then grown for four to five weeks using the roller tube technique (Gähwiler, 1981, 1984). Detailed protocols describing the culturing of CNS explants are provided elsewhere (Østergaard *et al.*, 1990, 1991). Immediately before processing, coverslips are removed from the roller tube and either fixed as complete cultures or snap-frozen on dry ice and sectioned on a cryostat (14 μm), thaw-mounted onto gelatin-coated coverslips and processed for ISH as standard (Finsen *et al.*, 1992) (see also Chapter 8).

10.3 Hybridization

We have found that the activity of these enzyme-conjugated oligonucleotides may be significantly affected by a multitude of variables, the quality of deionized distilled water being critical for optimal results. To avoid unnecessary "variables" we always use deionized double-distilled Milli-Q water for all stages of hybridization. Contamination of the water or buffers by heavy metal ions or related "gremlins" can significantly affect the strength of the resulting AP hybridization signal, to the point when no signal at all will be detected! All Milli-Q water is treated with the RNase inhibitor DEPC, as described in Chapter 1, and autoclaved before use. We tend to err on the side of caution and treat most buffers with DEPC; as mentioned in Chapter 1, Section 1.5.4, such cautionary steps may not be necessary but old habits die hard!

10.3.1 Estimating Optimal Hybridization and Washing Conditions

For oligonucleotides, optimal conditions, for example hybridization and washing temperatures, may be calculated empirically and are dependent upon the length of the probe, the GC content, the Na^+ concentration and the percentage of deionized formamide used (Chapter 1, Section 1.8.3). To determine the optimal conditions, it is important to determine the temperature at which 50% of DNA–RNA duplexes dissociate (referred to as the T_m). An estimation of the T_m for each oligonucleotide can be determined using the following basic equation (Wilkinson, 1992):

$$T_m\ (^\circ C) = 79.8 + 18.5\ \log(\text{molarity of monovalent cations}) + 0.58(\%\ G{+}C) + 0.12(\%\ G{+}C)^2 - 0.5(\%\ \text{formamide}) - (820 \div \text{probe length in bp})$$

PROTOCOL 10.3 STANDARD HYBRIDIZATION PROTOCOL FOR FRESH FROZEN SECTIONS

1. Remove sections from –80°C freezer and dry with a hairdryer.
2. Fix, dehydrate and air-dry as in Chapter 1.
3. Remove hybridization buffer from –20°C freezer and keep on ice. Hybridization buffer[a] contains: 50% deionized formamide,[b] 10% dextran sulfate, 4 × SSC, 1 × Denhardt's soln[c] and 250 μg/ml sonicated salmon testis DNA[d] (dithiothreitol is not necessary as no ^{35}S is present; it also inhibits AP activity).
4. Dilute AP-oligonucleotide in hybridization buffer. *Do not boil the probe.* Vortex *well*, return to ice and allow bubbles to clear before pipetting 250 μl of diluted probe on to each slide. Spread the buffer carefully over the sections using an "upturned" yellow Gilson pipette tip, being careful not to scratch the tissue.
5. Place plastic trays in humidified plastic bags and hybridize overnight (12–36 h, but no longer) at 37°C.

[a] This is a compromise between the "minimalist" and "maximalist" hybridization buffers described in Chapter 1; large volumes (40 ml) can be made up and stored at –20°C.
[b] It is imperative to maintain the activity of the AP at all times so it is recommended that the formamide be deionized using Amberlite M-80 resin (Sigma). Radioactive oligonucleotides are less sensitive to stray contaminations, such as metal ions, than AP probes. Mix 100 ml of formamide with 5 g of resin and stir gently at room temperature for 1 h in a sterile beaker. Filter the deionized formamide twice and store in 20-ml aliquots at –20°C.
[c] For convenience use RNase-free Denhardt's solution (Sigma D-2532) and
[d] RNase-free DNA (Sigma D-9156) and store frozen. Where a tissue has high endogenous AP activity treat with 0.2M HCl after fixation (see Table 10.1).

This equation should be used as a guide only, as it applies to DNA–RNA hybrids formed in solution and it is important to realize that the hybrids formed during ISH on fixed tissue are less stable and will therefore have an actual melting temperature slightly lower than the theoretical value calculated using the above formula.

The optimal incubation temperature for annealing is 10–15°C below the T_m for 50% formamide; for AP-oligonucleotides the true value of T_m is approximately 10°C lower than the empirical value.

10.3.2 HYBRIDIZATION TO FRESH FROZEN SECTIONS

In this section, variations on the standard radioactive ISH protocol (Chapter 1, Protocols 1.5 and 1.6) will be described and explanations offered for the modifications. The method that we have found to work well for most

AP-oligonucleotides is given in Protocol 10.3. In brief, fresh frozen sections are taken from the –80°C freezer just before the experiment, warmed to room temperature, then fixed and dehydrated as described in Chapter 1, air-dried and overlaid with hybridization buffer containing the diluted AP-oligonucleotide. It is important to dilute the AP-oligonucleotide in hybridization buffer just before application (see Section 10.2). Routinely we cover sections with 250 µl of buffer per slide (containing three to four sections). AP-oligonucleotides are diluted to give a final saturating probe concentration of 6–12 fmol/µl (1–2 µl/ml of 2 nmol in 200 µl of stock). This probe concentration is saturating for most neuropeptide, transmitter and receptor transcripts in CNS tissue but may need to be increased for detection of very abundant transcripts, for example actin mRNA. It is important to use saturating probe concentrations to ensure that there is sufficient probe to bind to all available hybridization sites; this criterion *must* be satisfied for semiquantitative analysis of ISH data (see Section 10.5). Sections are hybridized overnight at 37°C in humidified plastic/Perspex trays (Nunc catalogue no. 166508). The trays are sealed inside a plastic bag containing several "water soggy" paper towels and placed in an incubator/oven humidified by pots of distilled water. It is important that the sections do not dry out at all during this and any subsequent steps as this will lead to an increase in background staining and the "dried out" portion of the section(s) turning purple/black during the subsequent colour-development stage.

PROTOCOL 10.4 USING PARAFFIN-EMBEDDED SECTIONS.

1. Heat a hotplate with a flat surface to >60°C.
2. Take slide containing the mounted sections and quickly place on hotplate (section side up) until the paraffin wax melts (wax surrounding the embedded tissue will become transparent). This should take approx. 90 s for a section of rat CNS tissue.
3. Quickly, before wax re-sets dissolve wax in xylene (2 × 5 min).
4. Partially rehydrate sections, 100%, 90%, 70% ethanol (2 × 5 min each).
5. Rinse in 0.1 M PBS (10 min).
6. Overlay sections with 0.02% pepsin in 0.2 M HCl and incubate at 37°C for 30 min.
7. Refix sections with 4% paraformaldehyde in 0.1 M PBS (pH 7.4) for 30 min.
8. Rinse sections in 0.2% glycine in 0.1 M PBS (3 × 5 min).
9. Finally sections may be dehydrated, air-dried and overlaid with hybridization buffer as described in Protocol 10.3.

10.3.3 Hybridization to Paraffin-embedded Sections

There are numerous methods (Hoefler *et al.*, 1986; Wolber and Lloyd, 1988; Brahic and Haase, 1989; Pringle *et al.*, 1989) for localizing mRNA within paraffin-embedded tissue; many are variations on a "standard" protocol. Assuming that the tissue has been fixed, processed and sectioned to yield relatively intact mRNA, it is important to include in the protocol a proteinase step which allows the oligonucleotide access to the target transcript. The method that we have found to be successful for a variety of applications (for combining with antibody and radioactive ISH studies, for example) is given in Protocol 10.4.

10.4 Post-hybridization Washing and Colour Development

For AP-labelled oligonucleotides (26–36mers), we routinely wash hybridized sections according to a standard protocol: one room-temperature rinse in 1 × SSC followed by three washes in 1 × SSC at 55°C (30 min each). Higher temperature washes, especially in the presence of lauryl sulfate (SDS), are not advisable as this can result in sections floating off the gelatinized slides as well as in inhibition of AP enzyme activity. As noted in Chapter 1 (Section 1.8.4), it is probably the stringency of the hybridization that is critical in determining the specificity of the final hybridization signal. If the washing temperature is increased too much, the T_m of the oligonucleotide will be reached and all the specific DNA–RNA hybrids will melt. It is important to remember that AP-labelled oligonucleotides have an actual *Tm* some 10°C below the empirical value (see Section 10.3.1). After the final 1 × SSC 55°C wash, sections are rinsed in 1 × SSC at room temperature, then rinsed in a Tris buffer before being overlaid with AP substrate solution and incubated overnight at 25°C in the dark. Again, especially during the winter months, it is important that this stage is carried out in a humidified incubator; room-temperature incubations are fine in warmer climates (21–26°C). A recipe detailing the washing conditions and subsequent colour development is given in Protocol 10.5.

The intensity of the hybridization signal can be monitored by viewing slides under the light microscope (do not let sections dry out). When neuropeptide mRNAs in CNS tissue are being visualized, colour-development times can vary from 4 to 36 h, depending on the abundance of the mRNA. Sections may be dehydrated rapidly using acetone; however, dehydrating with absolute alcohol will significantly reduce the intensity of the AP hybridization signal as the reaction product is alcohol soluble.

PROTOCOL 10.5 POST-HYBRIDIZATION WASHING AND COLOUR DEVELOPMENT

1. Remove hybridized sections from the incubator, collect an aliquot of the discarded hybridization buffer[a] and rinse sections briefly in 1 × SSC at room temperature.
2. In a water bath at 55°C, wash sections three times with 1 × SSC, 30 min each wash.
3. Allow sections to cool in 1 × SSC (approx. 30–60 min), then rinse in Buffer A[b] (pH 7.4) for 30 min, then freshly prepared Buffer B[c] (pH 9.4) for 5 min.
4. During the Buffer B rinse, prepare the AP substrate solution; for a standard 26 mm × 76 mm slide, 0.7 ml of substrate solution per slide is sufficient. From the number of slides calculate the total volume of substrate solution required, then immediately before use add 4.5 µl of NBT stock[d] and 3.5 µl of BCIP stock[e] per ml of Buffer B required.
5. Remove slides, one at a time, from the Buffer B wash, lay them horizontally in incubation trays and overlay them with 0.7 ml of substrate solution. Ensure the tray is kept level at all times during the colour-development stage. *Do not let the sections dry out during this stage* as this will result in non-specific staining.
6. Finally, place slides in the dark at 25°C and leave the colour reaction to proceed.
7. The colour reaction can be terminated by washing sections in Stop buffer.[f]
8. Coverslip sections using glycerin jelly and store at 4°C to prevent gradual fading of the AP reaction product.

[a] Collect an aliquot of "used" hybridization buffer containing the AP probe in an Eppendorf tube and during the colour-development stage add approx. 0.5 ml of AP substrate solution to this solution. The solution in the Eppendorf tube should go dark blue within 1–2 h This test tells you that the substrate solution is OK, the pH of Buffer B is OK, and that the AP probe was added in excess. If this test tube does not go blue overnight you are unlikely to see any specific AP hybridization signal on the tissue sections.

[b] Buffer A: 100 mM Tris–HCl + 150 mM NaCl: pH 7.4.

[c] Buffer B: 100 mM Tris–HCl + 150 mM NaCl + 50 mM $MgCl_2$: pH 9.4.

[d] To make up a stock NBT (nitroblue tetrazolium) solution dissolve 75 mg of NBT chloride (Boehringer-Mannheim 1087-479) in 1 ml of 70% dimethylformamide; prepare in a glass tube only. Store in the dark at –20°C.

[e] To make up a stock solution of BCIP (5-bromo-4-chloro-3-indolyl-phosphate) dissolve 50 mg of BCIP toluidine salt (Boehringer-Mannheim 760-994) in 1 ml of 100% dimethylformamide. Store as above.

[f] Stop buffer: 100 mM Tris–HCl + 20 mM ED'I'A(Na_2) + 150 mM NaCl: pH 7.0–7.4.

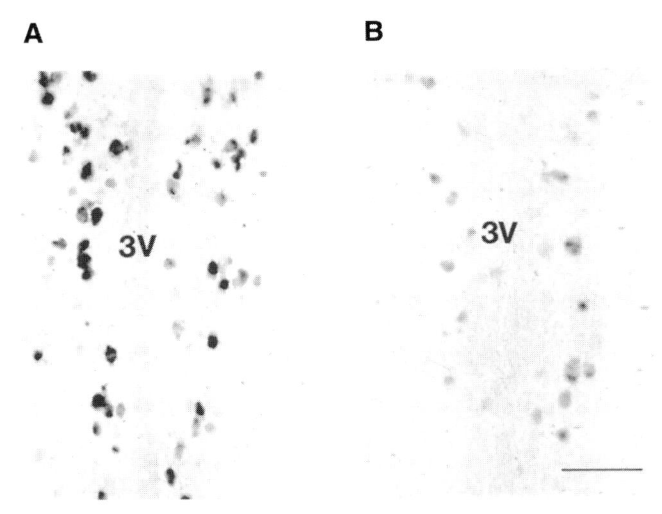

FIG. 10.3. The effect of differing Mg^{2+} concentrations on the intensity of the AP signal. (A) Detection of SRIF mRNA in the periventricular region of the rat hypothalamus using a final concentration of 50 mM Mg^{2+} in the colour reaction buffer. (B) a serial section of hypothalamus hybridized and developed as in (A) but with only 5 mM Mg^{2+} in the final buffer. Both the number of cells detected and the intensity of the signal is much greater when using buffer containing 50 mM Mg^{2+}. 3V, third ventricle. Scale bar 100 μm.

10.4.1 VARIABLES AFFECTING THE INTENSITY OF AP HYBRIDIZATION SIGNAL

Formamide

As mentioned earlier, formamide inhibits both calf intestinal (source of AP for the AP-oligonucleotides) and endogenous AP activity in rat. Determining the fine balance between a strong specific hybridization signal and a low background is important. Routinely, commercial producers of AP-oligonucleotides recommend that the hybridization buffer should contain only 20% formamide; however, such a low concentration can often result in high non-specific background staining. For membrane blots this may not be a problem but for ISH on tissue sections, high background staining can often "mask" the specific signal. To obtain the best signal/noise ratio, we always test all AP-oligonucleotides using a range of formamide concentrations in the hybridization buffer; usually 30–50% formamide yields the best results using our standard protocol. Post-hybridization sections are never washed in the presence of formamide.

pH of Buffer B and $MgCl_2$ Concentration

Both the pH and Mg^{2+} ion concentration of Buffer B are important. If the

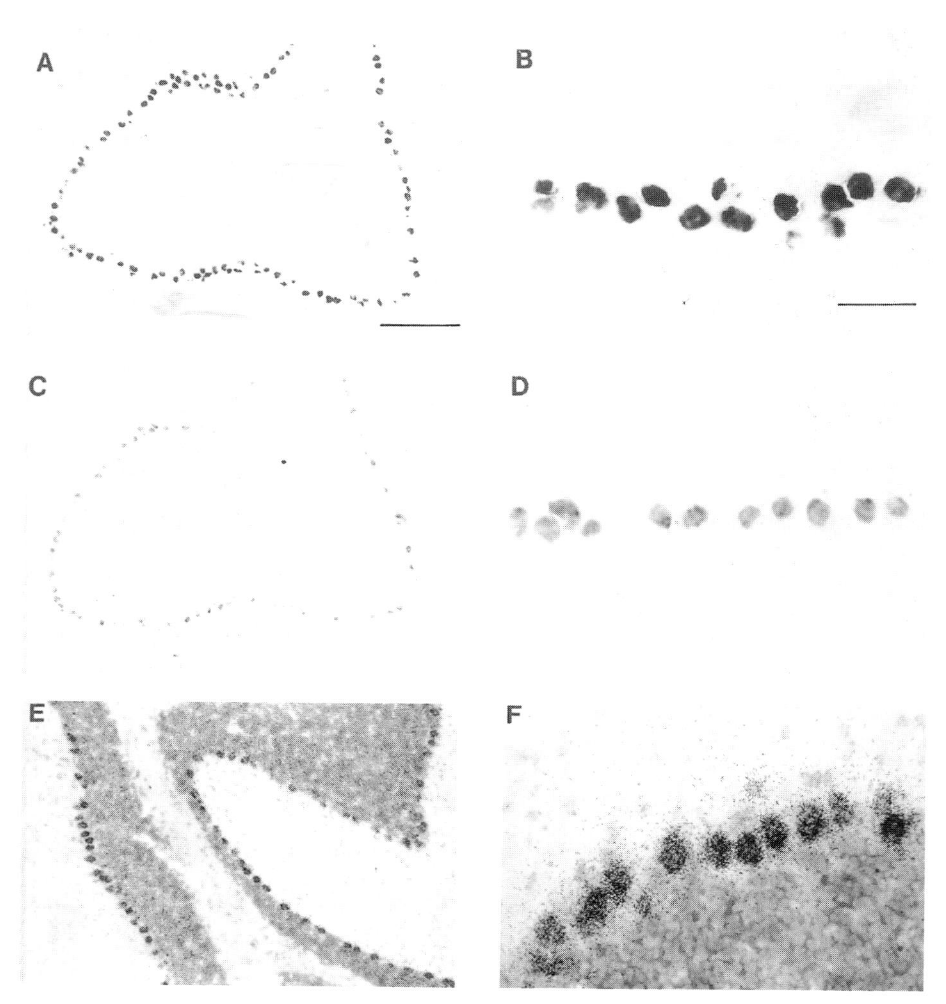

FIG. 10.4. Expression of calbindin-D28K in the Purkinje cells of the rat cerebellum using multiple oligonucleotides. (A) and (B) Detection of calbindin transcripts using four AP-conjugated probes (kindly supplied by British Biotechnology); compare the intensity of the signal with (C) and (D), where only one oligonucleotide was used; sections were processed in parallel. The resolution and intensity of the signal in (A) and (B) compare favourably with the detection of calbindin message using conventional radioactive ISH techniques with four [α-[^{35}S]thio]dATP-labelled calbindin oligonucleotides as shown in (E) and (F). A, C and E were taken at the same magnification; scale bar 100 μm. B, D and F are higher magnifications of the fields depicted in A, C and E respectively.

[Mg^{2+}] is too low or the final pH of Buffer B is not in the range 9.0–9.5, then the rate at which the colour reaction proceeds will be hindered and the intensity of the coloured AP reaction product, depicting sites of hybridization, will be very weak or possibly undetectable. From various pilot experiments, we have determined that the optimal [Mg^{2+}] is 50 mM. Figure 10.3 illustrates this point.

Number of Oligonucleotides

As with radioactive ISH (Chapter 1, Section 1.8.2 and Fig. 1.11), the intensity of the hybridization signal (and therefore the speed at which a signal is detected) may be markedly increased by hybridizing tissue sections with a mixture of several AP-oligonucleotides, each one being complementary to a different region of the transcript of interest. This approach is often adopted to visualize dopamine receptor mRNAs or other rare transcripts in brain using multiple radioactive oligonucleotides. Figure 10.4 demonstrates detection of calbindin mRNA in the rat cerebellum using either one or four AP-labelled oligonucleotides and compares this with four conventionally radiolabelled probes complementary to the same bases used in the design of the AP-labelled oligonucleotides.

10.5 Use of AP-oligonucleotides for Semiquantitive Analysis

If AP-oligonucleotides are to be used for semiquantitative analysis, for example as tools for assessing relative changes in the cellular content of an mRNA after pharmacological manipulation, it is important to (i) ensure that saturating probe concentrations are used, (ii) process all control and experimental sections together to ensure that all hybridization and colour-development conditions are standardized, and (iii) establish a colour-development time-course profile for each AP-oligonucleotide in the anatomical region of interest. An example of a colour-development time-course profile is illustrated in Fig. 10.5, showing that the intensity of the coloured AP hybridization signal increases linearly with time until 48 h, when the signal begins to plateau. Once the maximal intensity of signal is reached, the effective signal/noise ratio decreases as the intensity of the specific AP signal remains the same, whilst the level of background tissue staining increases with time. Colour-development times should not be extended beyond this point.

We have used AP-oligonucleotides for semiquantitative analysis to assess the relative changes in cellular neuropeptide/transmitter mRNA content

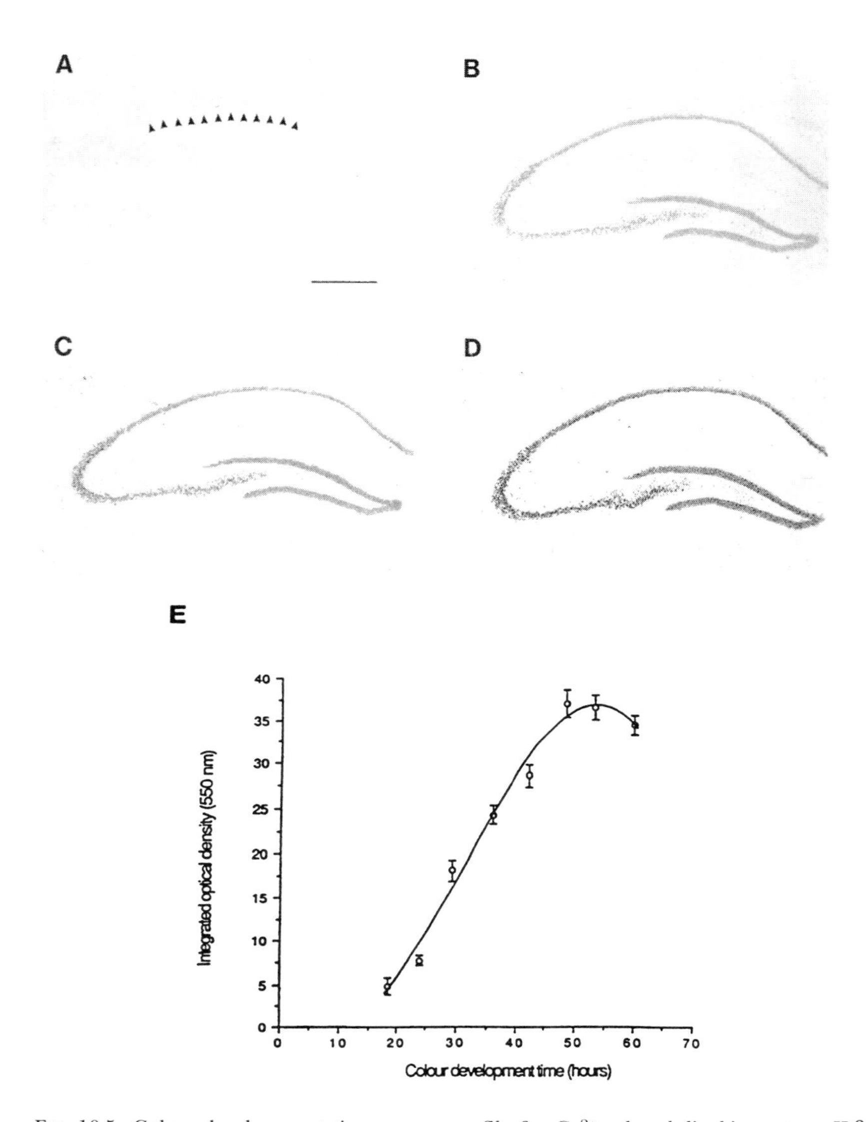

FIG. 10.5. Colour-development time-course profile for Ca^{2+} calmodulin kinase type II β (Cam-K II β) mRNA in the rodent hippocampus. The colour-development reaction was terminated at different time points, and the relative integrated optical density of the reaction product in the CA1 field of the hippocampus (indicated by arrows in A) was determined using a Vickers M85 microdensitometer. (A) 12-h development period, (B) 24-h development, (C) 36-h development, (D) 48-h development. (E) Graph showing the relationship between the density of the AP reaction product in the pyramidal cells of the hippocampus (OD) and the colour-development time. The intensity of the hybridization signal increases linearly until ~ 48 h after which it begins to plateau. Scale bar 250 μm.

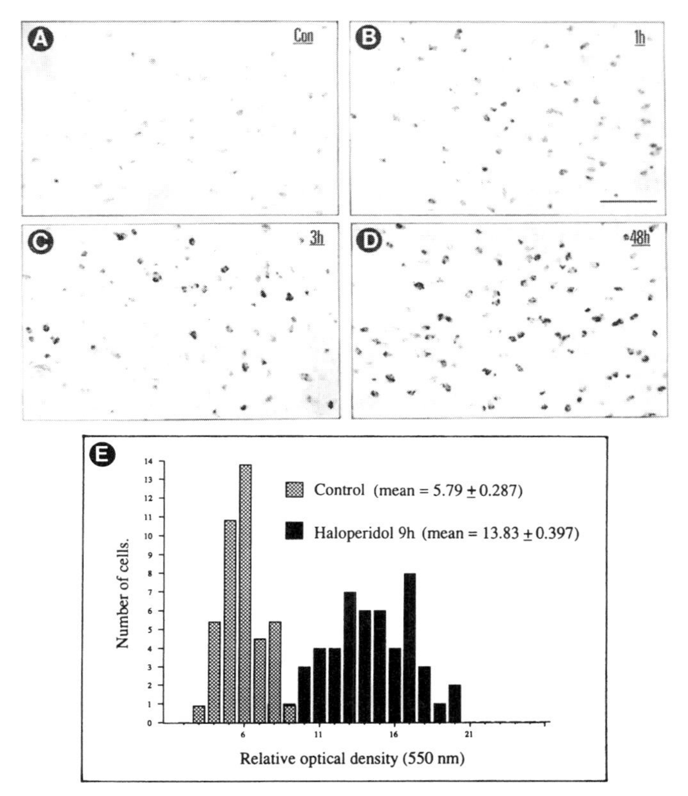

FIG. 10.6. Rapid and sustained increase in enkephalin gene expression in the rat striatum after administration of the neuroleptic, haloperidol (4 mg/kg, i.p.). (A) control, (B) 1 h, (C) 3 h, (D) 48 h after injection of the drug. Note the increase in the cellular content of enkephalin mRNA within striatal cells. The bar graph in (E) demonstrates that the increase in enkephalin gene expression is reflected by a shift to the right in the integrated optical density per cell, reflecting the increased cellular content of enkephalin mRNA. Scale bar 100 μm. Reproduced with kind permission from Elsevier Science Publishers.

after several pharmacological manipulations (Kiyama *et al.*, 1990b; Augood *et al.*, 1991a,b, 1992; Augood and Emson, 1992; Andreassen *et al.*, 1999). As can be seen in Fig. 10.6, an increase in the cellular content of preproenkephalin A mRNA in medium-sized striatal neurons after acute haloperidol administration is detected by the increased intensity of AP reaction product. It is important to note that all control and experimental sections are processed in parallel and that the colour-development time for all sections is the same.

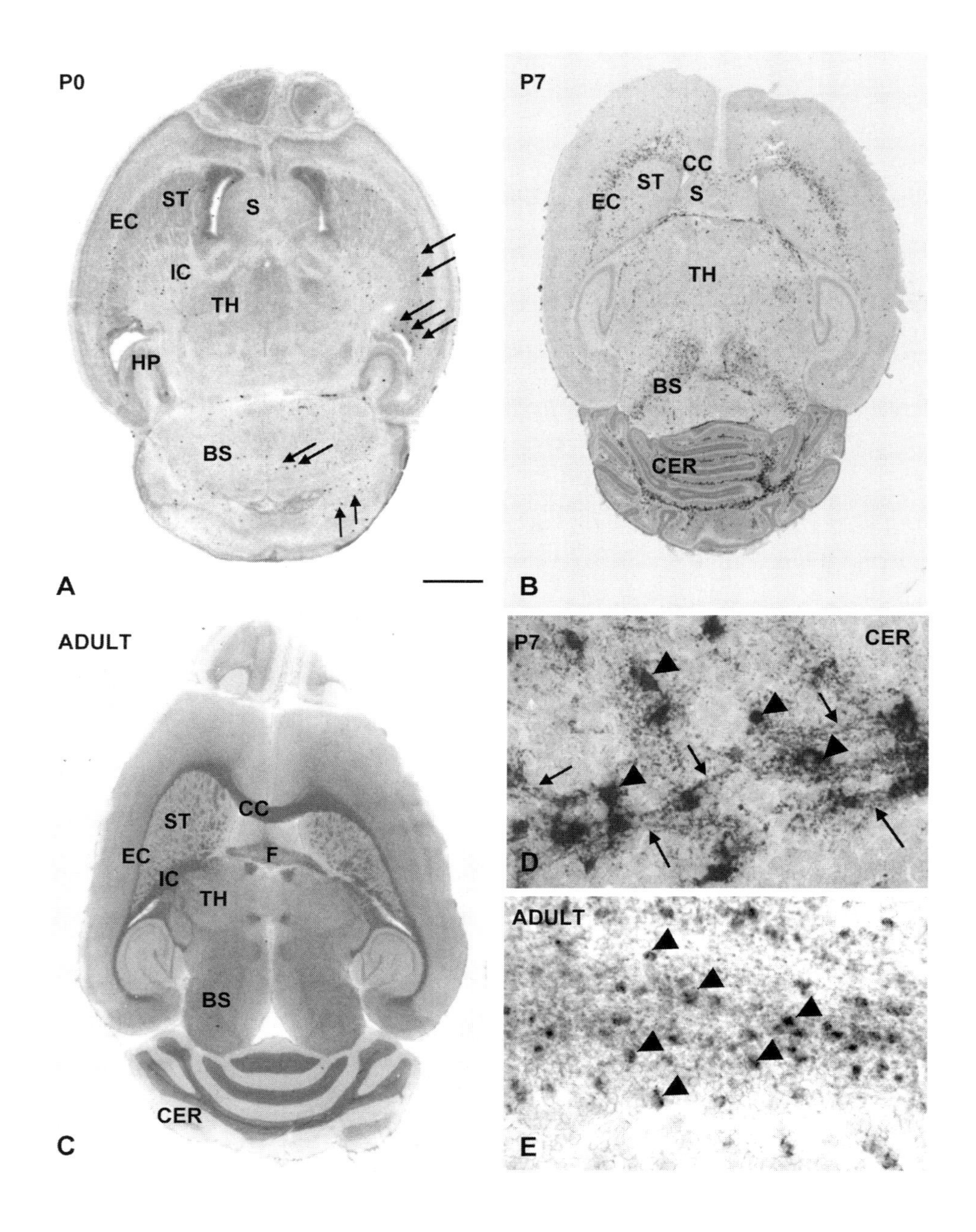
P0
ST
S
EC
IC
TH
HP
BS
A
P7
CC
ST
S
EC
TH
BS
CER
B
ADULT
CC
ST
EC
F
IC
TH
BS
CER
C
P7
CER
D
ADULT
E

Semiquantitative analysis of tissue sections hybridized with an AP-oligonucleotide may be carried out using a microdensitometer (Vickers M85) at 550 nm, the optimal wavelength for detecting AP reaction product using NBT and BCIP as substrates (Gutschmidt *et al.*, 1980). It is important to note that this type of analysis yields information about the relative changes in the *cellular* content of the mRNA of interest. This is unlike film autoradiography (used for the analysis of a radioactive hybridization signal), where changes in optical density readings for an area of interest indicate changes in the total mRNA content for that region. To establish if the changes in optical density measured off the autoradiograms reflect an increase in the cellular content of mRNA or/and an increase in the number of detectable cells expressing the mRNA of interest, sections must be processed for emulsion autoradiography, so that silver deposits can indicate, using light microscopy, the cellular sites of hybridization.

In situ hybridization with AP-labelled oligonucleotides thus represents a rapid means of detecting changes in mRNA expression, either due to experimental manipulations, or during physiological processes such as development (Fig. 10.7).

FIG. 10.7. MBP mRNA *in situ* hybridization. Developmental profile of oligodendrocyte myelin basic protein (MBP) mRNA expression in mouse brain studied with an alkaline phosphatase-tagged oligonucleotide. Horizontal brain sections of 16 μm (A,B,D) and 30 μm (C,E) were hybridized using a 30mer alkaline phosphatase-labelled DNA probe complementary to bases no. 302–333 of murine MBP cDNA (Newman *et al.*, 1987). The hybridized sections were lightly counterstained with haematoxylin, resulting in the neuronal staining that is observed in the cerebellum in B and C. (A–C) MBP mRNA is expressed in increasing numbers of oligodendrocytes from postnatal day 0 (arrows in A), through early postnatal development (B), to finally attain the expression levels observed in the adult CNS (C). Note in C the high MBP mRNA expression in the corpus callosum, the striatal grey matter, the external and the internal capsule, and in the cerebellar white matter of the adult brain. (D,E) High magnifications of cerebellar white matter in the P7 (D) and the adult (E) mouse. At both ages the diffuse hybridization signal between the strongly labelled oligodendrocyte cell bodies (arrowheads) correspond to MBP mRNA which is translocated into the processes of the oligodendrocytes. This is particularly evident in the P7 mouse (arrows). *In situ* hybridizations using the same probe as used in Jensen *et al.* (2000a,b). The *in situ* hybridizations were performed as described in Lambertsen *et al.* (2001). The probe was used at a concencentration of 2 pmol/ml. The probe was purchased from DNA Technology, Aarhus DK. BS, brain stem; CC, corpus callosum; CER, cerebellum; EC, external capsule; F, fornix; IC, internal capsule; HP, hippocampus; S, septum; ST, striatum; TH, thalamus. Scale bar: 1 mm (A).

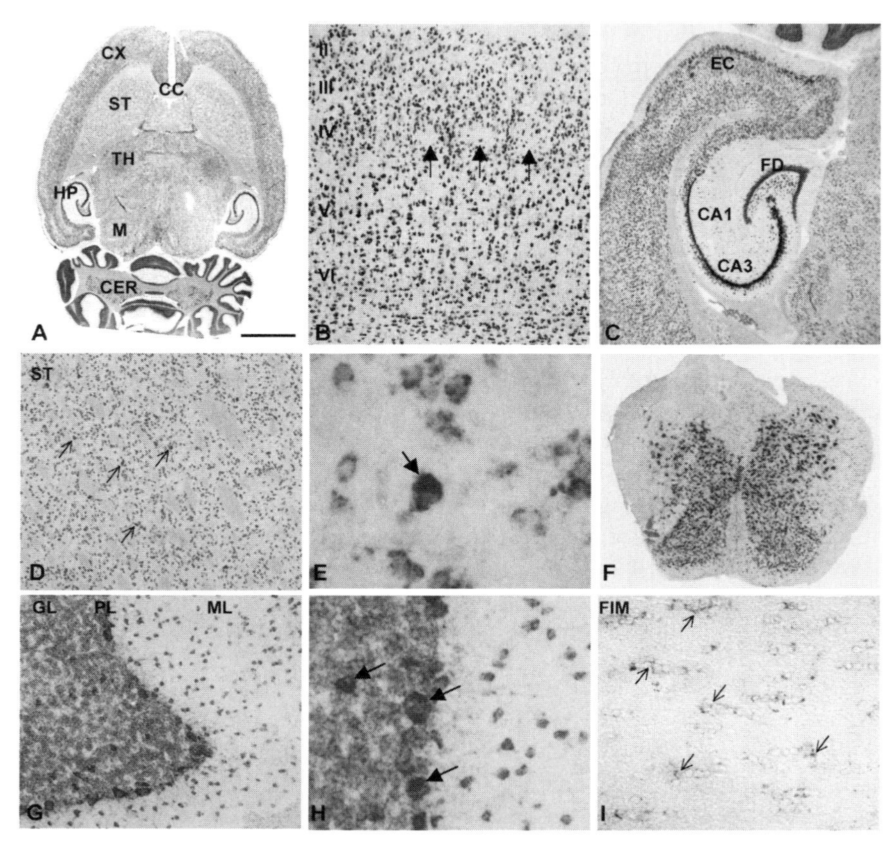

FIG. 10.8. GAPDH mRNA *in situ*. Regional and cellular expression of GAPDH mRNA in the adult CNS. Horizontal cryostat sections (16 μm) of rat (A–E,G,H) and of mouse (F, I) CNS subjected to *in situ* hybridization with 26mer alkaline phosphatase-tagged oligonucleotides complementary to bases no. 846–871 of rat (Fort *et al.* 1985) and bases no. 752–777 of mouse (Sabath *et al.* 1990) GAPDH cDNA. B–E show high magnifications from the section shown in A. (A) GAPDH mRNA is expressed at high levels in the neocortex, hippocampus, thalamus, mesencephalon and the cerebellum, and in low levels in the striatum and in the white matter. (B) GAPDH mRNA-expression pattern in rat somatosensory cortex. Note individual barrels (arrows) within cortical layer IV. (C) GAPDH mRNA expression in the hippocampal formation. GAPDH mRNA is confined to the neuronal somata. (D,E) With the exception of a few polymorphic neurons (arrows in D and arrow in E), neuronal GAPDH mRNA expression is less abundant in the striatum. E shows a high magnification of D. (F) Tranverse section of spinal cord from mouse outlining the difference in GAPDH mRNA expression between grey and white matter. (G,H) GAPDH mRNA expression in the cerebellum. H is a higher magnification of G visualizing the high GAPDH mRNA-expression level in the Purkinje cells and in the Golgi-like cells within the granular layer (arrows). (I) High-power photomicrograph of the fimbria revealing GAPDH mRNA expression in oligodendrocytes (arrows). The *in situ* hybridizations were performed as described in Gregersen *et al.* (2000) and Lambertsen *et al.*

10.6 Suitable Controls for ISH

Problems associated with determining the "ideal" ISH control are discussed in Chapter 1 (Section 1.10).

We routinely hybridize "control" sections with an excess of unlabelled oligonucleotide to demonstrate that the binding of the AP-oligonucleotide is displaceable. In addition, some sections are treated with RNase prior to hybridization stage to demonstrate that the probe is binding to single-stranded RNA (mRNA). Finally, northern analysis is occasionally carried out using the cold oligonucleotide labelled with [^{32}P]dATP. Hybridizing sections for GAPDH mRNA is a useful control for the quality of the ISH procedure and the quality of the sections (Fig. 10.8).

10.7 Combining AP-oligonucleotides and [^{35}S]Oligonucleotides: Co-expression Studies

For the simultaneous visualization of two mRNAs within a single neuron, we developed a technique of co-expression (Kiyama *et al.*, 1991; Augood *et al.*, 1993, 1995, 1997) in which one mRNA is visualized using an AP-oligonucleotide and the other using an ^{35}S-labelled oligonucleotide. Ideally the two oligonucleotides used should be of similar length and GC content, so that the conditions of stringency for the two probes are similar. The least-abundant mRNA should be detected using the radioactive probe; for example, for the localization of dopamine (DA) receptor mRNAs and neuropeptide mRNAs in the rat basal ganglia, the DA receptor oligonucleotide would be radiolabelled with α^{35}SdATP (see Chapter 1) and the neuropeptide oligonucleotide labelled with AP. Examples of co-expression are shown in Figs 10.9 and 10.10.

FIG. 10.8 *continued*

(2001). CA1, regio superior hippocampus; CA3, regio inferior hippocampus; CC, corpus callosum; CER, cerebellum; CX, neocortex; EC, entorhinal cortex; FD, fascia dentata; FIM, fimbria; GL, granule cell layer; HP, hippocampus; M, mesencephalon; ML, molecular layer; PL, Purkinje cell layer; ST, striatum; TH, thalamus. Scale bar: 400 μm (A), 250 μm (B,D); 1000 μm (C), 30 μm (E), 500 μm (F), 125 μm (G), 40 μm (H) and 50 μm (I)

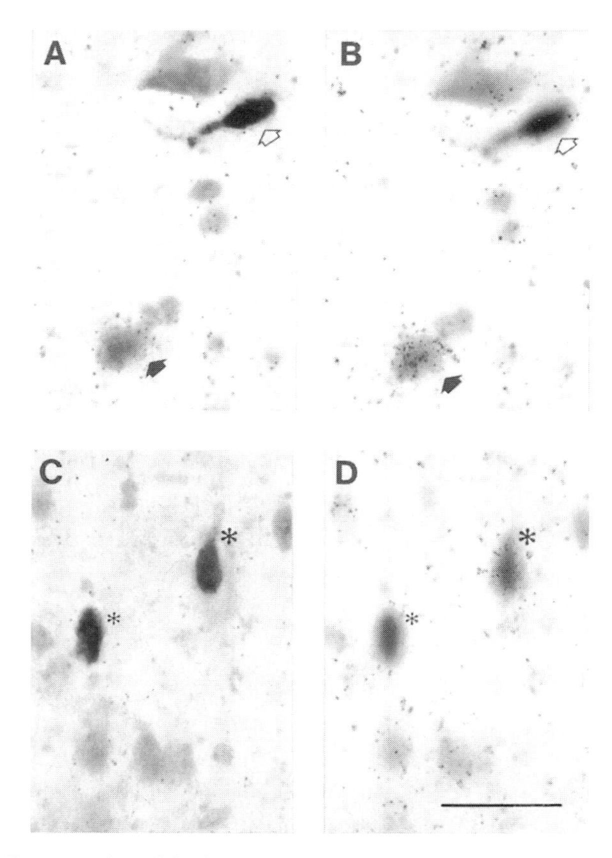

FIG. 10.9. Co-expression of the AMPA receptor subunit, GluR2 mRNA and SRIF mRNA in the cerebral cortex. As noted in the text, clear and accurate photographic representation of co-expression studies can be difficult as the silver grains and AP reaction product are in different planes of focus. This is illustrated in (B) and (D), and (A) and (C), respectively. The same cortical field of view focused in (A) on the AP reaction product (SRIF mRNA) and (B) on the silver grains (GluR2 mRNA), the neuron strongly expressing SRIF does not contain any GluR2 transcripts (open arrow) and the GluR2-expressing cell contains no AP reaction product (closed arrow); the faint staining in the GluR2-expressing neuron is non-specific background. Similarly with (C) focused on AP reaction product (SRIF mRNA) and (D) focused on silver grains (GluR2 mRNA), it can clearly be seen that one cortical neuron (large asterisk) contains both transcripts, and the other neuron (small asterisk) expresses only SRIF. Scale bar 100 μm.

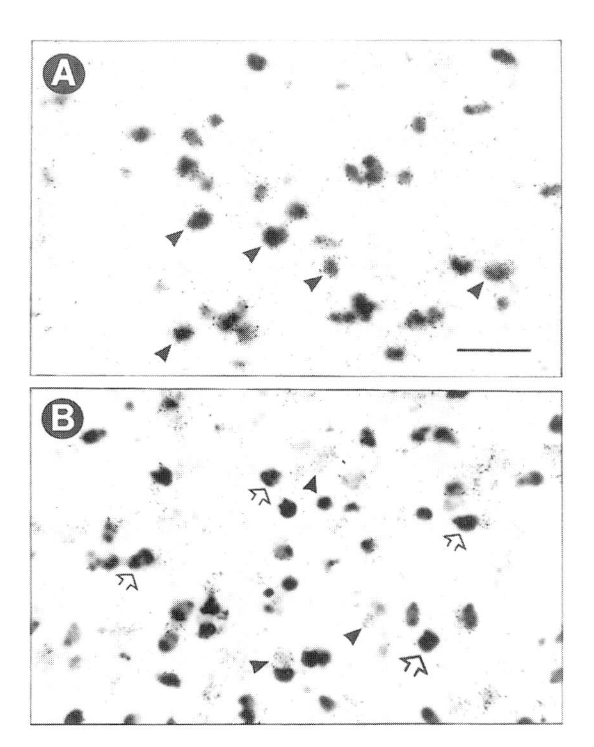

FIG. 10.10. Cryostat sections of rat striatum hybridized simultaneously with (A) AP-enkephalin (ENK) probe and ^{35}S-labelled dopamine D_2 receptor oligonucleotide (silver grains), and (B) AP-ENK oligonucleotide and ^{35}S-labelled dopamine D_1 receptor oligonucleotide (silver grains). (A) Both hybridization signals (purple AP reaction product and silver grains) are concentrated within the same cells, indicating that ENK-expressing cells contain dopamine D_2 mRNA. Conversely in (B) the two hybridization signals appear concentrated in different cells illustrating that, in control rats, ENK (filled arrowheads) and dopamine D_1 receptor (open arrows) transcripts are not generally co-expressed. Scale bar 50 μm. (Reproduced with kind permission from Elsevier Science Publishers.)

We have successfully demonstrated the cellular co-expression of two mRNAs in both fresh frozen cryostat sections and paraffin-embedded material (Kiyama *et al.*, 1991; Augood *et al.*, 1993; Emson *et al.*, 1993), illustrating further the versatility of this approach. Essentially, the standard hybridization protocol is followed (see Protocol 10.3 for fresh frozen sections and Protocol 10.4 for paraffin-embedded tissue) except that both the AP- and ^{35}S-labelled probes are diluted in the same hybridization buffer (with the addition of 3 μl/ml β-mercaptoethanol) and applied simultaneously to the fixed tissue section. Sections are then washed in 1 × SSC at 55°C and processed for colour development as detailed in Protocol 10.5. Once the AP reaction product is seen in the cytoplasm of positive cells, sections are washed extensively in Stop buffer, then rapidly dehydrated through a graded series of alcohol or acetone and processed for autoradiography (using Ilford K5 emulsion). For details of emulsion autoradiography refer to Chapter 1 (Protocol 1.9). After the desired exposure time to emulsion (routinely 4–14 weeks for neuropeptide/receptor mRNAs), sections should be developed and coverslipped with glycerin jelly. It is important to remember *not* to counterstain the sections. Using a light microscope the two hybridization signals can be resolved easily; the coloured reaction product (corresponding to the AP-oligonucleotide) will be seen concentrated within the cell cytoplasm, and the silver grains, depicting sites where the ^{35}S-labelled oligonucleotide has hybridized, will be seen clustered over cells in a different plane of focus (see Fig. 10.7). After colour development, to prevent non-specific discoloration of the tissue sections during the emulsion autoradiography stage, sections must be washed several times in Stop buffer. To aid visualization of both the radioactive and AP hybridization signals, we have found it best to stop the AP colour reaction before the mRNA-containing cells exhibit an intense colour signal, as this can sometimes mask the silver-grain labelling, i.e. it is difficult to see black silver grains clustered over a cell containing a dark purple/black-coloured reaction product. The reverse is also true: the silver grain labelling must not be so intense that the AP reaction product concentrated "underneath" in the cell cytoplasm cannot be clearly seen. A compromise of hybridization signal intensity for both probes is ideal.

10.8 Combining AP ISH and Immunohistochemistry

Although we have successfully combined AP ISH with immunohistochemistry on fresh frozen cryostat sections, paraffin-embedded material is the tissue of choice because of the excellent preservation of tissue morphology. As noted earlier, the order in which the two techniques are carried out can be critical if a good signal is to be obtained for both the ISH and immunocytochemistry. Although we generally perform the ISH step first and the immunohistochemistry second, there are examples when the immunohistochemistry must be performed first, otherwise the antigen will be lost or denatured to such an extent (presumably by the formamide in the hybridization buffer) that it is not recognized by the antibody. This problem has been commented on in the literature by several authors (Schalling *et al.*,

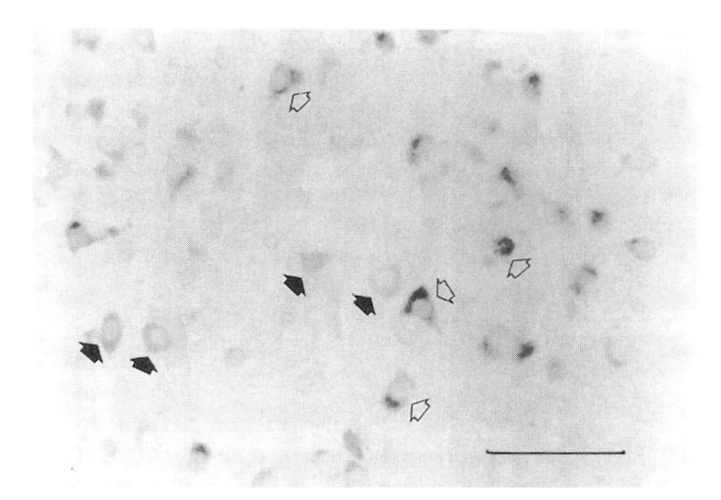

FIG. 10.11. Colour photomicrograph demonstrating the co-localization of calretinin mRNA (purple reaction product) and calbindin-D28K immunoreactivity (brown diaminobenzidine product) in a paraffin-embedded section of rat substantia nigra. Most neurons in the substantia nigra contain calbindin-D28K (closed arrows) and a subpopulation of these neurons also express calretinin mRNA (open arrows). Scale bar 75 μm. This figure is reproduced in colour between pages 202 and 203

1986; Shivers *et al.*, 1986; Watts and Swanson, 1989); indeed, for localization of oestrogen receptor immunoreactivity in the nucleus of hypothalamic cells, the antibody step must be carried out first using RNase-free (DEPC-treated) solutions.

The protocol we use routinely involves carrying out the ISH step first (see Protocol 10.4), then, after detection of the AP reaction product, sections are rinsed extensively in Stop buffer and finally processed as detailed in Protocol 10.6. Examples of the high quality of tissue morphology that can be obtained using paraffin-embedded tissue is given in Fig. 10.1 and in Fig. 10.11, which show the co-localization of calretinin mRNA and calbindin immunoreactivity in cells of the substantia nigra.

PROTOCOL 10.6 COMBINING ISH AND IMMUNOCYTOCHEMISTRY

(It is not necessary to use RNase-free solutions for all the stages given below.)

1. First process sections for ISH using an AP-oligonucleotide (see Protocol 10.3 or 10.4).
2. After visualization of the AP reaction product, rinse tissue well in Stop buffer.
3. Rinse tissue for 10 min in 0.1 M PBS.
4. Incubate sections in 5% NGS[a] + 0.5% Triton X-100 in 0.1 M PBS to block non-specific binding.
5. Incubate sections in primary antibody diluted in PBS + 1% normal serum + 0.5% Triton X-100 for 24–48 h at 4°C. Sections may need to be overlaid with parafilm to prevent them drying out.
6. Rinse sections for 5 × 10 min in PBS.
7. Incubate sections in horseradish peroxidase-conjugated goat anti-rabbit IgG[b] (1:500 for 2 h or 1:1000 overnight at 4°C). Alternatively use a Vector ABC kit and follow the manufacturer's instructions.
8. Wash sections for 5 × 10 min in PBS, then react with 0.05% DAB[c] activated with H_2O_2 (add 12 µl of 33% H_2O_2 to 100 ml of DAB solution).
9. Finally rinse sections well in an excess of PBS and coverslip using glycerin jelly.

[a] NGS, normal goat serum Vary the species depending on the host of the IgG.
[b] Assuming the primary antibody was raised in rabbit, supplied by Vector.
[c] DAB, diaminobenzidine. This solution should be made up immediately before use.

10.9 Troubleshooting

Table 10.1 describes some problems encountered in AP hybridization and gives some reasons and remedies.

Acknowledgements

The authors would like to thank the MRC for financial support. N. Tønder, K. Westmore and I. King are thanked for their excellent technical assistance.

TABLE 10.1
PROBLEMS ENCOUNTERED IN AP HYBRIDIZATION.

Problem	*Suggested reason and remedy*
High AP background	Tissue sections dried out at some stage. High endogenous AP activity, e.g. in the intestine; treat sections with 0.2 M HCl for 10 min after fixation (see Protocol 10.3). AP-oligonucleotide too concentrated. Reduce [probe]. BCIP stock soln should be clear, if coloured (brown) make up fresh stock.
No colour reaction	Was AP probe added to HYB buffer? Check purity of ddH_2O; Milli-Q H_2O preferred for colour reaction. Check $MgCl_2$ concentration of Buffer B. Check that pH of Buffer B is between pH 9.0 and 9.5. Check activity of AP-oligonucleotide stock by spotting 0.5μl of neat probe on a piece of filter paper, then add a drop of AP substrate solution (containing NBT and BCIP: see Protocol 10.5). The spot should go blue/purple within 1–3 min. No colour reaction suggests very low/no AP enzyme activity. Trying to detect a rare transcript; technique not sensitive enough. Try 3′-end-labelling oligonucleotide with [α-[^{35}S]thio]dATP or use riboprobes.
Yellow crystals	Check temperature of colour development (20–25°C). Microfilter AP substrate solution just before use.
Low signal	Check through above pointers. Signal may be increased by mixing together several AP-oligonucleotides that are complementary to different regions of the same transcript of interest. Signal-to-noise ratio may be increased by reducing the formamide concentration in the hybridization buffer from 50% to 40% or even 30% (to compensate for the inhibition of AP activity by formamide). Note of caution, this reduces the hybridization stringency.
High background after emulsion dipping	Sections not washed well enough in Stop buffer.

References

Agrawal, S., Christodoulou, C., and Gait, M. J. (1986). *Nucleic Acids Res.* **14,** 6227–6245.

Andreassen, O. A., Finsen, B., Østergaard, K., Sorensen, J. C., West, M. J., and Jorgensen, H. A. (1999). *Neuroscience* **88,** 27–35.

Augood, S. J. and Emson, P. C. (1992). *Neuroscience* **47,** 317–324.

Augood, S. J., Kiyama, H., Faull, R. L. M., and Emson, P. C. (199la). *Mol. Brain Res.* **9,** 341–346.

Augood, S. J., Kiyama, H., Faull, R. L. M., and Emson, P. C. (199lb). *Neuroscience* **44,** 35–44.

Augood, S. J., Faull, R. L. M., and Emson, P. C. (1992). *Eur. J. Neurosci.* **4,** 102–112.
Augood, S. J., Westmore, K., Faull, R. L. M., and Emson, P. C. (1993). *Mol. Brain Res.* **20,** 328–334.
Augood, S. J., Herbison, A. E., and Emson, P. C. (1995). *J. Neurosci.* **15,** 865–874.
Augood, S. J., Westmore, K., and Emson, P. C. (1997). *Neuroscience* **76,** 763–774.
Brahic, M. and Haase, A. T. (1989). *Curr. Top. Microbiol. Immunol.* **143,** 9–20.
Emson, P. C. (1993). *Trends Neurosci.* **16,** 9–16.
Emson, P. C., Heppelmann, B., and Augood, S. J. (1993). *In* "*In situ* Hybridization", second edition (J. D. Bardhas, K. Valentino, and J. Eberwine, Eds), pp. 43–62. Oxford University Press, Oxford.
Finsen, B. R., Tønder, N., Augood, S. J., and Zimmer, J. (1992). *Neuroscience* **47,** 105–113.
Fort, P., Marty, L., Piechaczyk, M., Sabrouty, S., Dani, C., Jeanteur, P., and Blanchard, J. M. (1985). *Nucleic Acids Res.* **13,** 1431–1442.
Gähwiler, B. H. (1981). *J. Neurosci. Meth.* **4,** 329–342.
Gähwiler, B. H. (1984). *Neuroscience* **11,** 751–760.
Gregersen, R., Lambertsen, K., and Finsen, B. (2000). *J. Cereb. Blood Flow Metab.* **20,** 53–65.
Gregersen, R., Christensen, T., Lehrmann, E., Diemer, N. H., and Finsen, B. (2001). *Exp. Brain Res.* **138,** 384–392.
Gutschmidt, S., Lang, U., and Riecken, E. O. (1980). *Histochemistry* **69,** 189–202.
Heppelmann, B., Señaris, R., and Emson, P. C. (1994). *Brain Res.* **635,** 293–299.
Hoefler, H., Childers, H., Montminy, M. R., Lechan, R. M., Goodman, R. H., and Wolfe, H. J. (1986). *Histochem. J.* **18,** 597–604.
Hopman, A. H. N., Wiegant, J., Tesser, G. I., and Van Duijin, P. (1986). *Nucleic Acids Res.* **14,** 6471–6488.
Hougaard, D. M., Hansen, H., and Larsson, L. I. (1997). *Histochem. Cell Biol.* **108,** 335–344.
Jablonski, E., Moomaw, E. W., Tullis, R. H., and Ruth J. L. (1986). *Nucleic Acids Res.* **14,** 6115–6128.
Jensen, M. B., Hegelund, I. V., Lomholt, N. D., Finsen B., and Owens, T. (2000a) *J. Neurosci.* **20,** 3612–3621.
Jensen, M. B., Poulsen, F. R., and Finsen, B. (2000b). *Int. J. Dev. Neurosci.* **18,** 221–235.
Kadowaki, K., McGowan, E. M., Mock, G., Chandler, S., and Emson, P. C. (1993). *Neurosci. Lett.* **153,** 80–84.
Kiyama, H. and Emson, P. C. (1990). *Neuroscience* **38,** 223–244.
Kiyama, H., Emson, P. C., and Ruth, J. (1990a). *Eur. J. Neurosci.* **2,** 512–524.
Kiyama, H., Emson, P. C., Ruth, J., and Morgan, C. (1990b). *Mol. Brain Res.* **7,** 213–219.
Kiyama, H., McGowan, E., and Emson P. C. (1991). *Mol. Brain Res.* **9,** 87–93.
Lambertsen, K. L., Gregersen, R., Drojdahl, N., Owens, T., and Finsen, B. (2001). *Brain Res. Protocols* **7,** 175–191.
Langer, P. R., Waldrop, A. A., and Ward, D. C. (1981). *Proc. Natl Acad. Sci. USA* **78,** 6633–6637.
Newman, S., Kitamura, K., and Campagnoni, A. T. (1987). *Proc. Natl Acad. Sci. USA* **84,** 886–890.
Ohnuma, T., Augood, S. J., Arai, H., McKenna, P. J., and Emson, P. C. (1999). *Neuroscience* **93,** 441–448.
Østergaard, K., Schou, J. P., and Zimmer, J. (1990). *Exp. Brain Res.* **82,** 547–565.
Østergaard, K., Schou, J. P., Gähwiler, B. H., and Zimmer, J. (1991). *Exp. Brain Res.* **83,** 357–365.
Østergaard, K., Finsen, B., and Zimmer, J. (1995). *Exp. Brain Res.* **103,** 70–84.
Pedersen, K. M., Finsen, B., Celis, J. E., and Jensen, N. A. (1998). *J. Biol. Chem.* **273,** 31494–31504.
Pringle, J. H., Primrose, L., Kind, C. N., Talbot, I. C., and Launder, I. (1989). *J. Pathol.* **158,** 279–286.
Sabath, D. E., Broome, H. E., and Prystowsky, M. B. (1990). *Gene* **91,** 185–191.

Schalling, M., Hökfelt, T., Wallace, B., Goldstein, M., Filer, D., Yamin, C., and Schlesinger, D. H. (1986). *Proc. Natl Acad. Sci. USA* **83,** 6208–6212.

Shivers, B. D., Harlan, R. E., Pfaff, D. W., and Schachter, B. S. (1986). *J. Histochem. Cytochem.* **34,** 39–43.

Watts, A. G. and Swanson, L. W. (1989). *In* "Methods in Neurosciences", Vol. 1 (P. M. Conn, Ed.), pp. 127–136. Academic Press, London.

West, M. J., Østergaard, K., Andreassen, O. A., and Finsen, B. (1996). *J. Comp. Neurol.* **370,** 11–22.

Wilkinson, D. G. (1992). *In* "*In situ* Hybridization, a Practical Approach" (D. G. Wilkinson, Ed.), pp. 1–13. IRL Press, Oxford.

Wolber, R. A. and Lloyd, R. V. (1988). *Hum. Pathol.* **19,** 741–763.

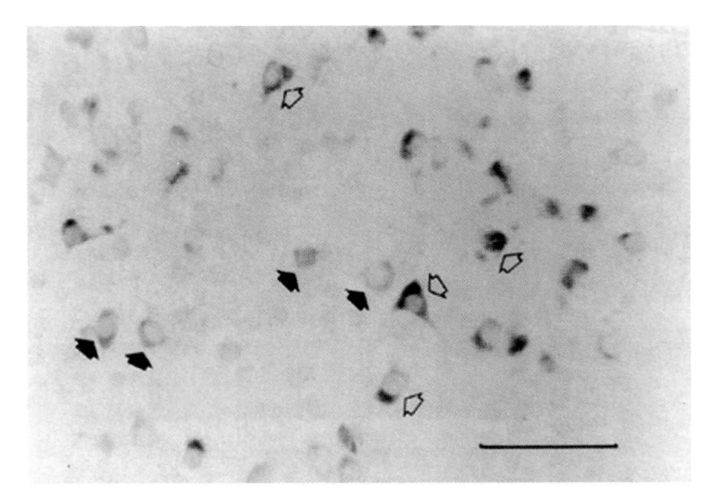

FIG. 10.11. Colour photomicrograph demonstrating the co-localization of calretinin mRNA (purple reaction product) and calbindin-D28K immunoreactivity (brown diaminobenzidine product) in a paraffin-embedded section of rat substantia nigra. Most neurons in the substantia nigra contain calbindin-D28K (closed arrows) and a subpopulation of these neurons also express calretinin mRNA (open arrows). Scale bar 75 μm. This figure is reproduced in colour between pages 202 and 203

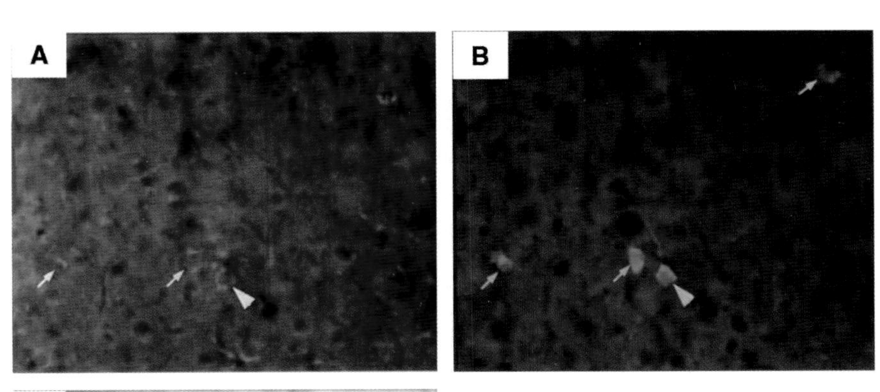

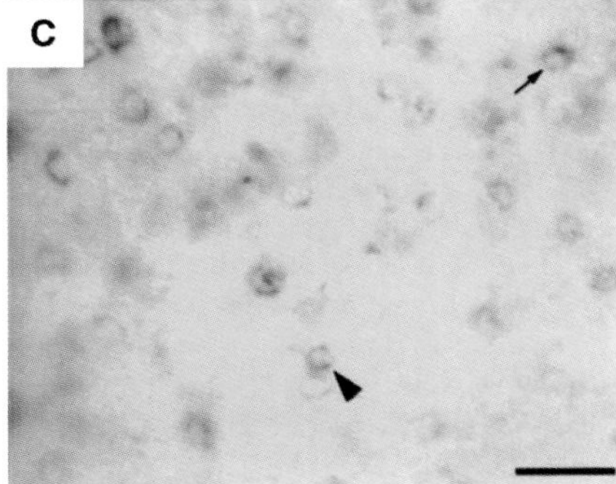

FIG. 11.5. Parvalbumin-immunoreactivity neurons expressing trkB or trkC mRNA are partially overlapping subsets. Triple labelling for trkB mRNA (A), parvalbumin immunofluorescence (B) and trkC mRNA (C) in visual cortex sections *in vivo*. Large arrowheads indicate a triple-labelled neuron, small arrows double-labelled neurons. Scale bar in C 50 mm for all photomicrographs. (Reproduced from Gorba and Wahle, 1999). This figure is reproduced in colour between pages 202 and 203.

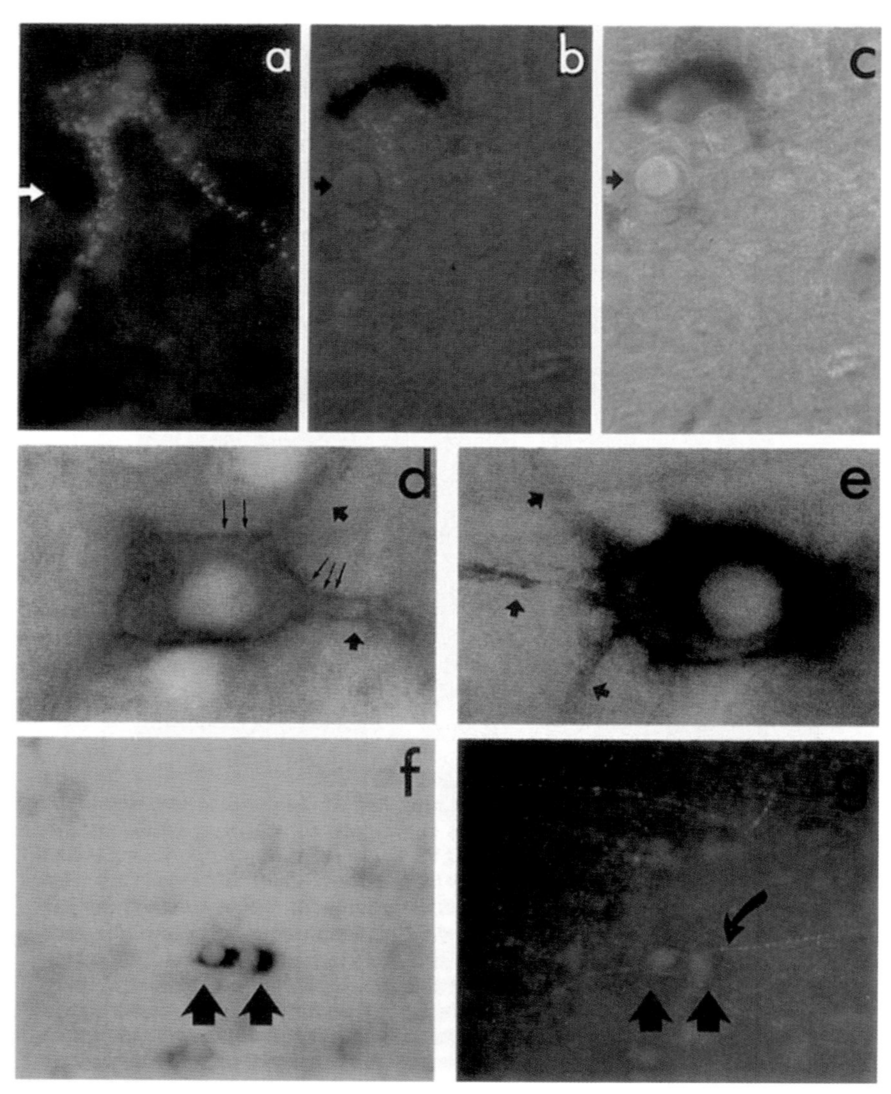

FIG. 11.4. Double-labelling experiments. (A) Retrogradely transported rhodamin-conjugated latex beads visualized with fluorescence have accumulated in the soma and dendrites of a neuron in the cat's nucleus of the optic tract. (B) GAD mRNA was visualized, and both products can be seen in a mixed exposure. (C) Bright-field exposure shows the AP reaction product in this neuron. (D) GAD mRNA, red reaction product, in a VVA lectin-labelled neuron. (E) Same as in D but blue reaction product. Neurons were from adult cat visual cortex. Fat arrows point to labelled processes; the fine arrows indicate the typical fenestrated VVA-labelling pattern. (F, G) Organotypic culture (explanted at postnatal day 2 and maintained for 2 weeks *in vitro*) of rat visual cortex displays NPY mRNA-expressing neurons (F, the bold arrows), which colocalize NPY-immunoreactive material, as revealed with FITC-labelled secondary antibody (G). The curved arrow points to the axon of one cell. Note the many immunoreactive beaded processes and puncta indicating the dense NPY innervation of the culture. This figure is reproduced in colour between pages 202 and 203.

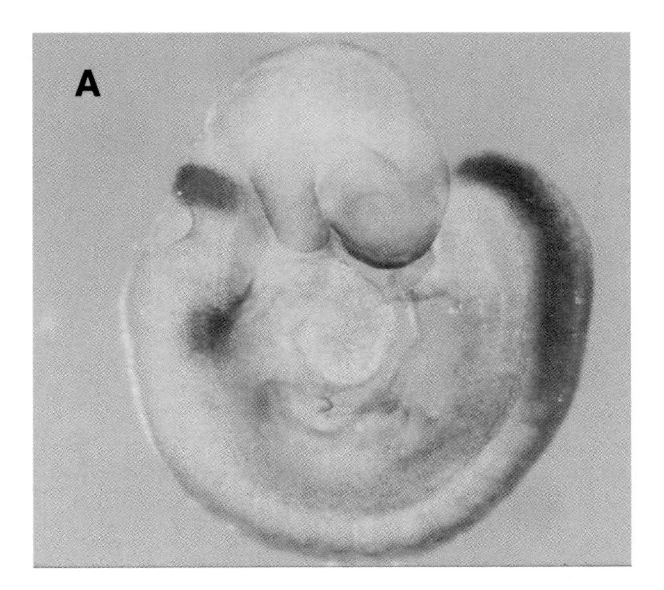

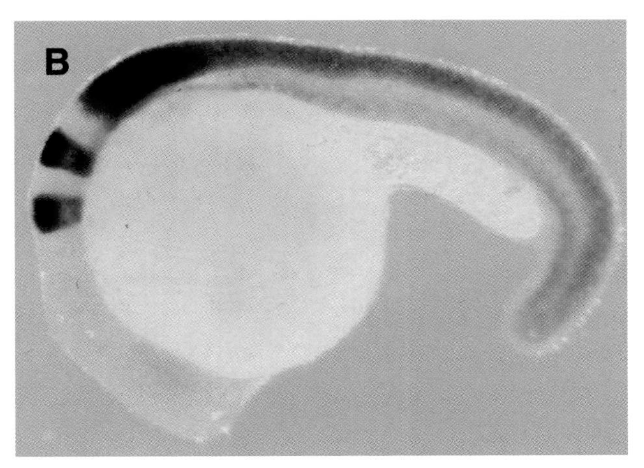

FIG 12.1. Patterns of expression of genes in two different species. (A) Whole-mount *in situ* hybridization of 9.5 dpc mouse embryo hybridized with a Hoxb1 RNA probe. There is a band of expression in the hindbrain corresponding to rhombomere 4. At this stage the expression in the neural tube has regressed posteriorly. (B) Zebrafish embryo (20 h), where two different probes have been used simultaneously. The two stripes correspond to Krox20, which is expressed in the hindbrain in rhombomeres 3 and 5. The most posterior staining in the neural tube corresponds to the pattern of expression of a presumptive Hoxa4 gene. This figure is reproduced in colour between pages 202 and 203.

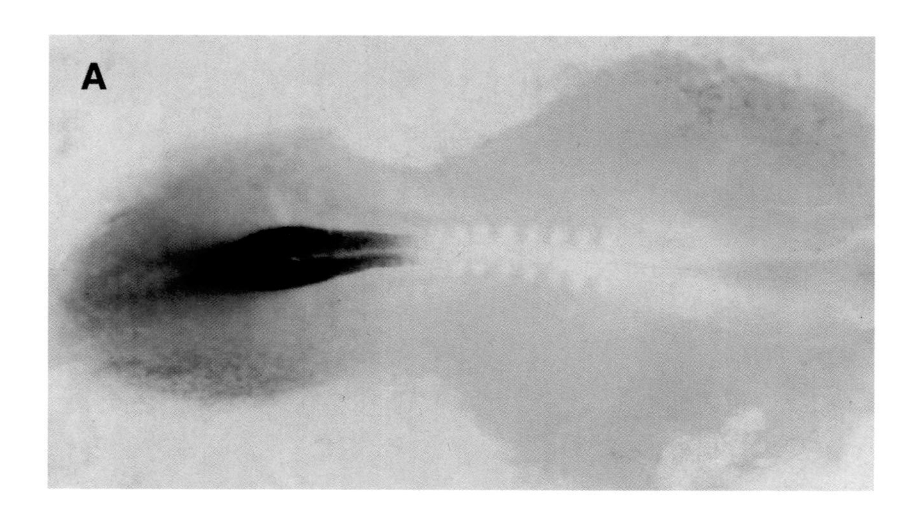

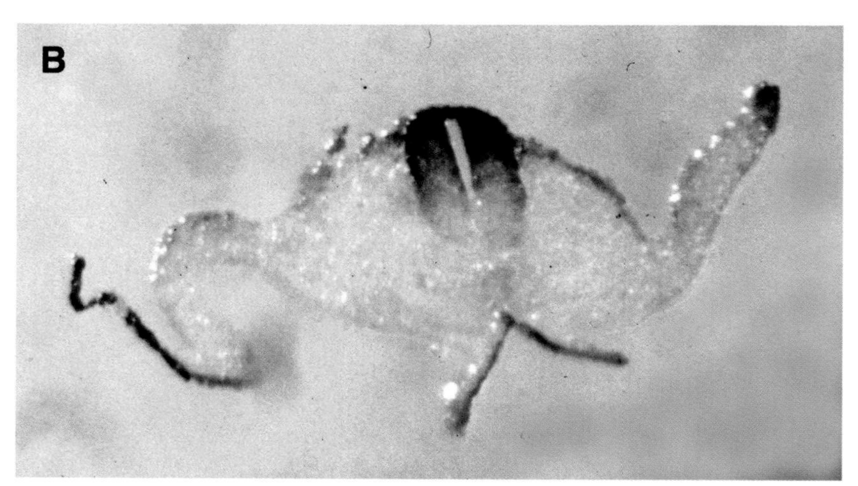

FIG. 12.2. Whole-mount *in situ* hybridization of a chick embryo. (A) Dorsal view of stage 8+ embryo, with chick Hoxb9 RNA probe. At this stage, there is a strong posterior expression in the presomitic mesoderm and neural tube. (B) Cryostat transverse section of similar embryo processed with the chick Hoxb9 RNA probe. Sectioning shows us that expression is stronger in the dorsal than in the ventral region. This figure is reproduced in colour between pages 202 and 203.

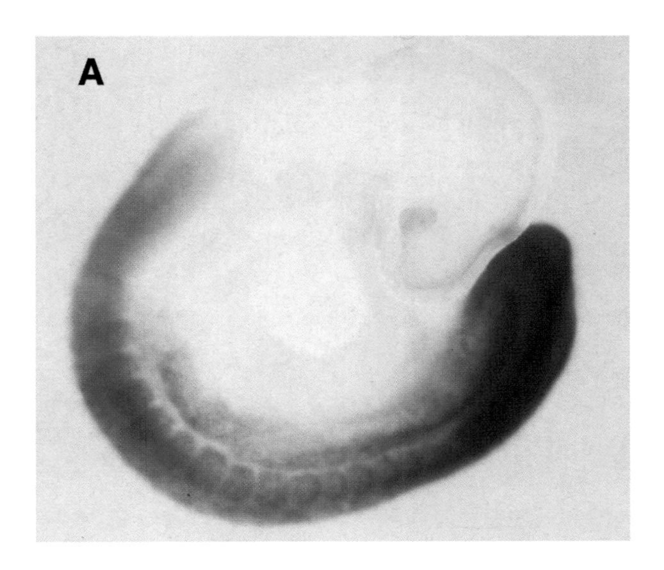

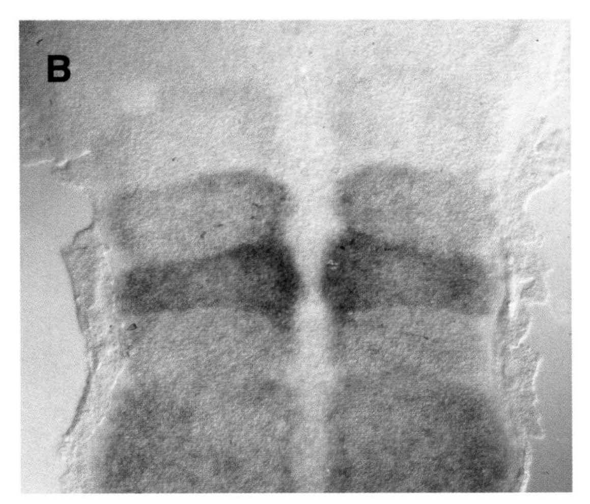

FIG. 12.3. Embryos hybridized with different hox genes. (A) 9.5 dpc mouse embryo hybridized with Hoxd4 gene, which is expressed along the neural tube and somites. Anterior expression is limited to rhombomere 6/7. (B) Flat-mounted hindbrain of a chick embryo stage 15, hybridized with the chick Hoxb2 probe. A clearer pattern of expression can be seen by dissecting out the tissue of interest and mounting it on a microscope slide. This figure is reproduced in colour between pages 202 and 203.

CHAPTER 11

COMBINING NON-RADIOACTIVE *IN SITU* HYBRIDIZATION WITH IMMUNOHISTOLOGICAL AND ANATOMICAL TECHNIQUES.

P. Wahle

AG Entwicklungsneurobiologie, Fakultät für Biologie, Ruhr-Universität, ND 6/72, D-44780 Bochum, Germany

INTERNATIONAL REVIEW OF NEUROBIOLOGY, VOL. 47
ISSN/02 $00.00

11.1 Introduction

11.1.1 Why do we Need to Identify mRNA-expressing Cell Types?

In situ hybridization allows us to localize in any given tissue the cells that express the mRNA for a particular gene. Because of the structural complexity of nervous tissue, it is of great importance to identify the neuronal (or glial) cell types expressing the mRNA because this will deliver information on the functional circuits into which the mRNA-expressing cell type is integrated. This information becomes very important in the light of accumulating evidence that not only the functional, but also the molecular, features of neurons are maintained and can be modified by activity-dependent mechanisms. Analysis of gene expression in a complex neuronal network thus requires the localization of mRNA and the concurrent characterization of molecular, morphological and anatomical features of the labelled cells. Furthermore, we are also analysing the developing CNS, dealing with very fragile tissue. A method is needed that preserves its structural integrity. Any *in situ* hybridization protocol employed must therefore meet these requirements.

11.1.2 Why Use Non-radioactive cRNA Probes?

This chapter describes a method of *in situ* hybridization using digoxigenin (DIG)-UTP-labelled cRNA probes (see Chapter 12 for the application of this method to embryos as whole-mount preparations; and appendix A9.38 of Sambrook and Russell (2001) for a detailed background on the digoxigenin system applied to labelling nucleic acid probes). We use this method for both brain sections and organotypic cultures (e.g. Obst and Wahle, 1995, 1997; Gorba and Wahle, 1999; Gorba *et al.*, 1999; Heumann *et al.*, 2000; Wahle *et al.*, 2000).

The DIG molecule is a unique steroid isolated from the plant *Digitalis purpurea.* Linked to UTP, it can be incorporated as a marker into cRNA probes. Specific antibodies against DIG then allow the detection of the DIG-labelled hybrid RNAs, and the expressing cells are stained with a colorimetric reaction product.

The reasons for employing non-radioactive methods have been discussed before (see Emson, 1993, and Chapter 10). A big advantage of using non-radioactive methods is they allow double or triple labelling of cells. For example, two different mRNAs can be co-localized in the same cell (Section 11.6.3; Gorba and Wahle, 1999); protein and mRNA can be co-localized (section 11.6.1; Heumann *et al.*, 2000); mRNA localization can be combined with tract tracing/retrograde labelling (Section 11.6.4; Wahle *et al.,* 1994; Schmidt *et al.,* 1998, 2001); and two mRNAs plus protein can be co-localized (Gorba and Wahle, 1999).

Other advantages of non-radioactive probes are that there is no need for isotopes, which are expensive in purchasing and in disposal; radioactive waste is not produced and there is no need for a special laboratory with extra equipment for making and storing radioactive probes, performing experiments and developing the material (isotope lab, darkrooms). There is also the absence of radiation risk for personnel. The non-radioactive hybridization techniques are particularly useful for teaching students in university labs. Even naïve beginners are able to run an experiment with positive outcome within the limited time of an introductory lab course. Furthermore, DIG-labelled probes are stable for months, since they have no physically decaying tag. cRNA probes, however, are synthetic RNA and thus more sensitive to degradation than DNA oligonucleotides. To protect the probe, gloves must be worn throughout the hybridization, and equipment has to be as clean as possible or RNase-free. Once this becomes routine, however, riboprobes are at least as easy to handle as antibodies.

An *in vitro* transcription yields enough probe for a series of experiments. Entire masters courses have been run in my lab with probes from one transcription reaction. This is advantageous for a developmental study, for example, because the variable "probe activity" can be excluded. Further, the RNA polymerases synthesizing the probes are highly specific for their promoter, generating either antisense (complementary to cellular mRNA) or, a suitable control, sense (the same as the cellular mRNA) probes. Non-radioactive probes are detected immunologically. The respective methods are straightforward and standard in many labs, as are enzyme-based colorimetric detections. The procedure is short and can be carried out within 2–3 days. The signal development can be controlled by eye, using a microscope, and the hybridized tissue can be directly processed for detection of additional markers. For a discussion of quantification of digoxigenin-based signals, see Chapter 9, Section 9.6.

11.1.3 Organization of the Chapter

This chapter is organized in the order in which an experiment would be carried out, starting with probe preparation, tissue preparation, the hybridization and subsequent detection of cell type-specific markers. At the end of the sections the outcome and staining results are discussed. At the end of each section, recipes are given for the buffers and solutions, and a list of the equipment used in our lab. Molecular biology techniques not directly related to the topic of the article may be acquired from "Current Protocols in Molecular Biology" (Ausubel *et al.*, 1987 and recent updates) or "Molecular Cloning: A Laboratory Manual; 3rd edition" (Sambrook and Russell, 2001).

11.2 The Probes

11.2.1 Preparation of Plasmid-DNA for Transcription

DNA fragments of interest are inserted into plasmids/transcription vectors (Ausubel *et al.*, 1987, chapter 1.8; Sambrook and Russell, 2001). To prepare the cRNA probe, template DNA (e.g. obtained by RT-PCR from brain mRNA) is cloned into a plasmid (transcription vector). In our lab we isolate brain mRNA using a Dynabead mRNA direct kit (Dynal, Hamburg, Germany); cDNA is synthesized using avian myeloblastosis virus (AMV) reverse transcriptase (Stratagene) and PCR done with GoldStar DNA-polymerase (Eurogentec, Belgium) or Taq polymerase (Qiagen, Heidelberg). The PCR primers contain restriction enzyme sites for directional cloning into pBluescript (Gorba and Wahle, 1999). Once a clone has been obtained and characterized, plasmids are grown. We previously used 100-ml cultures (Ausubel *et al.*, 1987, chapter 1.7) and isolated the plasmid DNA with anion exchange columns (e.g. Qiagen Pack-100). The kit comes with the buffers, a detailed protocol, and yields a large amount of pure plasmid DNA. We recently found that normal miniprep plasmid DNA can be successfully used for transcriptions, and therefore switched to growing only 10-ml cultures, from which plasmids are isolated with alkaline lysis. The bacterial RNA is degraded with RNase A. The DNA is aliquoted and stored as an ethanol pellet.

About 5–10 μg of the DNA is digested with restriction enzymes cutting downstream of the promoter and insert (Ausubel *et al.*, 1987, chapter 3). Linearization is checked on an agarose gel. DNA from minipreps are then purified with at least three phenol–chloroform (1 : 1) extractions, followed by at least three chloroform extractions, ethanol-precipitated overnight in the freezer (Ausubel *et al.*, 1987, chapter 2; Sambrook and Russell 2001, appendix A8.9), pelleted by centrifugation, washed three times with 80% ethanol (made with DEPC-treated water), and dried with a SpeedVac. Gloves, sterile equipment and RNA-grade chemicals are used.

PROTOCOL 11.1 *IN VITRO* TRANSCRIPTION

1. To 1 μg template dissolved in 13 μl water on ice add
   2 μl 10× transcription buffer
   2 μl 10× nucleic acid labelling mixture
   2 μl (20–40 units) RNA polymerase
   1 μl RNase inhibitor.
2. Vortex briefly and spin down to bottom of the tube.
3. Incubate at 37°C for 2 h.
4. Add 10–20 units of an RNase-free DNase to digest the template DNA and mix (this is an optional step).
5. Incubate at 37°C for 15 min.
6. Add 1 μl 0.5 M EDTA pH 8.0 (DEPC-treated!) to stop the reaction. Mix.
7. Precipitate the labelled RNA with 1/10 vol. 4 M LiCl and 3 vol. ice-cold 100% ethanol.
8. Leave at –20°C overnight.
9. Pellet the RNA with centrifugation at 14 000 g for 30 min.
10. Pour off supernatant, wash pellet with 50 μl 80% ethanol. A large brownish pellet, often somewhat fuzzy, should be obtained.
11. Remove the ethanol with a pipette.
12. Cover tube opening and inner lid with piece of parafilm (the unexposed side). Poke in a tiny hole with a sterile needle.
13. Dry the pellet in a SpeedVac. The parafilm protects the RNA from airborne contaminants.
14. Dissolve the pellet in 90 μl DEPC-treated water at 37°C for 30 min. Make sure water, tubes and pipette tips are RNase-free. Vortex or pipette up and down.
15. When dissolved, take 2 μl into a separate tube.
16. To the rest add 10 μl of a 10× TER (100 mM Tris–HCl pH 7.5, 50 mM EDTA pH 8.0, 1 mg/ml yeast tRNA). Mix. An RNase inhibitor may be added to 50 U/ml.
17. Store the riboprobe at –20°C.

The linearized DNA is dissolved in 20 μl of RNase-free water (pure double-distilled water; we use water from a Millipore unit, treated with DEPC, and autoclaved twice). Then 1 μl is run together with a standard DNA of known concentration on an agarose gel stained with ethidium bromide to determine the concentration (Ausubel *et al.*, 1987, chapter 2; Sambrook and Russell, 2001, appendix A8.23). The linearized template should appear as a single sharp band of the predicted size. About 1 μg in 10 μl water is used for the transcription reaction. It is allowed to stand for 30 min on ice to make sure that the DNA is completely dissolved (it is slowly pipetted up and down several times with sterile RNase-free pipette tips, or vortexed carefully) (Protocol 11.1, see also Chapter 12, Protocol 12.1).

11.2.2 *In Vitro* Transcription

Gloves, sterile equipment and RNA-grade chemicals must be used. Riboprobes are synthesized with the Nucleic Acid (RNA) Labeling kit (Roche). This contains a 10× transcription buffer, RNA polymerases (SP6/T7 or T3/T7 at 20 U/µl), and a 10× nucleic acid mixture containing the digoxigenin (DIG)-labelled UTP (10 mM ATP, CTP, GTP; 6.5 mM UTP and 3.5 mM DIG-UTP). Buffers and polymerases from other companies work well, too. An RNase inhibitor may be included in the reaction mixture, but is not necessary if reagents are RNase-free. The reaction volume is 20 µl.

11.2.3 How to Check Transcription Efficiency and Quality of the Probes

Newly synthesized probes should be checked by gel electrophoresis and on a northern blot (Protocol 11.2). Gloves, sterile equipment and RNA-grade chemicals should be used. To check the cRNA size, the DIG-labelled RNA of known size (760 bp) and concentration (100 ng/µl) that comes with the Roche RNA Labeling kit can be used. (Fig. 11.1).

The results are shown in Fig. 11.1. These cRNAs had been dissolved in 100 µl, and 1 µl was run on the gel. Transcripts from different templates are shown. The labelled cRNA should appear as more or less sharp bands. Highly concentrated probes yield overloaded lanes, and bands appear broad and somewhat smeary. Multiple bands may occur due to ineffective denaturing. Such probes can be used at high dilutions. The size of the transcripts can be determined using the control RNA band of 760 bp. The concentration can be judged by comparing the staining intensity of the bands. Less efficient transcriptions yield fainter bands. These probes can still be good for *in situ* hybridization, but have to be used at higher concentrations. A smear below a band would indicate partial transcripts. Such probes may still be used for ISH. Degraded probes may run as smears with no band of the expected size.

The absence of bands or a faint smear at the bromophenol–dye front indicates failure of transcription. Usually this can already be seen after precipitation of the transcription reaction (no pellet). Reasons may be impurities of template DNA, badly dissolved DNA, wrong concentration of buffer or RNase contamination of ingredients or equipment (see Ausubel *et al.* (1987) for troubleshooting the transcription reactions).

PROTOCOL 11.2 TEST RIBOPROBES BY NORTHERN BLOT

1. Take 1 µl of control RNA.
2. Heat-denature at 65°C for about 5 min (optional).
3. Mix it with an equal volume (1 µl) of methylmercury hydroxide (MeHgOH, from SERVA)-containing loading buffer (see Sambrook and Russell. 2001; chapter 7 on gel electrophoresis of RNA). Note, that MeHgOH is very toxic, so work in a fume hood.
4. Heat 1–2 µl of the newly synthesized DIG-labelled cRNA dissolved in water to denature, and mix with 2 µl MeHgOH. Store samples on ice.
5. Prepare a minigel. Take a glass plate (ours are 5 × 7.5 cm) or large glass slide that fits into a minigel chamber and a comb. Clean equipment with rinses in 1% SDS-water, followed by sterile water. Dry comb and glass plate by a rinse in technical grade ethanol. Mount comb 0.5–1 mm above the glass plate with two gel clamps on a level bench.
6. Dissolve RNA-grade agarose to 1% in borate buffer. Allow to cool to 55°C.
7. Add ethidium bromide to 0.2 µg/ml.
8. Pipette, with a baked glass pipette, 10 ml of 1% agarose on top of the glass plate. Surface tension keeps the solution on the glass, and it solidifies quickly. The comb forms shallow slots. Mount gel into chamber, fill with borate buffer.
9. Carefully load the RNA samples (the slots take up to 5 µl), and run the gel in at 50 V.
10. Visualize bands under UV light.
11. Perform a Northern blot (Ausubel *et al.*, 1987, chapter 4.9; Sambrook and Russell, 2001, chapter 7, protocol 7) with the gel.
12. Transfer in 20 × SSC for at least 4 h (better overnight when thicker standard minigels are used) onto nylon membrane (Hybond N, Amersham).
13. UV cross-link the RNA to the membrane (Stratagene cross-linker).

The following steps are not performed under sterile conditions. The buffer recipes are given at the end of the chapter.

14. Rinse the membrane 2 × 2 min in distilled water, then 2 × 5 min in buffer DIG-I.
15. Block with blocking reagent (Roche) diluted in DIG-I for 15 min at room temperature.
16. Briefly rinse off block solution and incubate membrane in anti-DIG antibody (sheep $F(ab)_2$ fragments; Roche) tagged with alkaline phosphatase, diluted 1 : 5000 in DIG-I for 1 h at room temperature.
17. Rinse in DIG-I for 3 × 5 min.
18. Equilibrate for 1–2 min in two changes of TBS-Mg buffer pH 9.5.
19. Incubate in substrate solution in TBS-Mg buffer using nitroblue tetrazolium (NBT) and X-phosphate (BCIP). For the incubations, membranes are heat-sealed into plastic bags. Volumes should be kept as low as possible (for a 5 × 7.5 cm membrane use 5 ml solution). Signals should develop within 30 min. Stop the reaction in water after the bands have appeared. Allow the filter to dry.

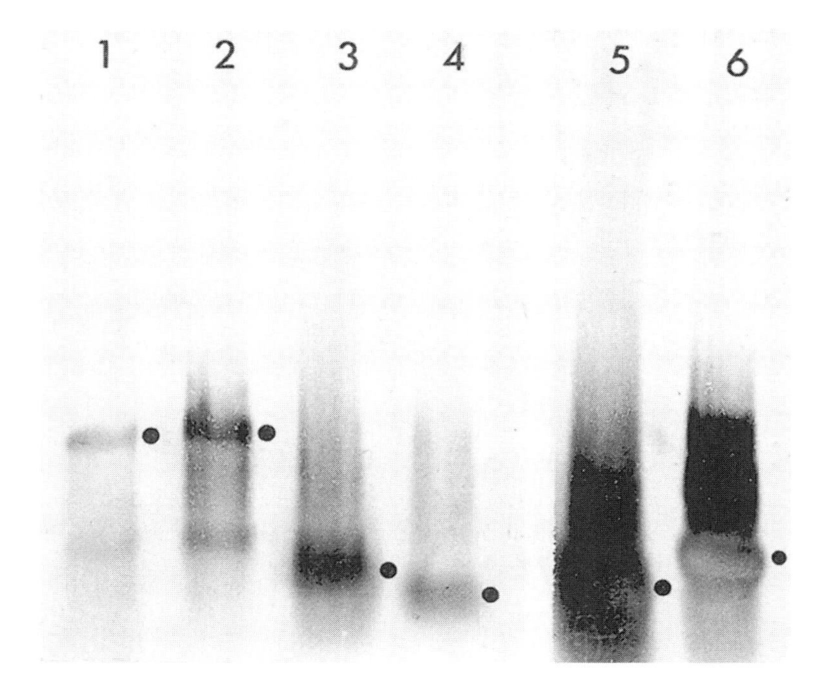

FIG. 11.1. Northern blot of a minigel reveals *in vitro* transcribed cRNA probes labelled with DIG-UTP. Lane 1, 2.3 kb GAD antisense cRNA; lane 2, 2.3 kb GAD sense cRNA; lane 3, 0.4 kb NPY exon 2 antisense; lane 4, 0.28 kb NPY signal peptide antisense; lane 5, 0.34 kb ß-preprotachykinin exons 1–4 antisense; lane 6, 0.76 kb DIG-labelled control RNA. Major bands are indicated by dots. Note that bands have different intensities as a result of different transcription efficiencies. The major band in lane 6 is so overloaded that, owing to steric hindrance of antibody, it appears less well stained than the rest. Lanes 1 and 2 have partial transcripts running below the expected cRNA size. This occurs sometimes, but has no effect on ISH results.

11.2.4 Storage of Probes

We store the labelled RNA at –20°C in aliquots of 20–30 µl. Repeated freeze–thaw does not change probe quality. Probes should be thawed in the hand. If RNase inhibitor has been included, do not thaw at 65°C, because this will inactivate the RNase inhibitor and may release bound RNases. Probes are vortexed before use. There should be no precipitate. We are currently using probes synthesized over eight months ago, and probe quality has remained unchanged. We usually divide good cRNA probes into small aliquots, which are stored long term as ethanol precipitate. Once an aliquot runs out, the next one is spun down and dissolved to the original volume.

PROTOCOL 11.3 EQUIPMENT AND SOLUTIONS FOR PROBES

Equipment: Gel chamber, plates, combs, clamps, power supply, nylon membrane (Amersham Hybond N), UV light table 302 nm, polaroid camera, microlitre pipettes, sterile Eppendorf tubes and tips.
Reagents: Agarose RNA-grade, technical ethanol, restriction enzymes (New England Biolabs, buffers included), blocking reagent and sheep-anti-DIG labelled with alkaline phosphatase ($F(ab)_2$ fragments, Roche).
Solutions:
Diethylpyrocarbonate (DEPC)-treated water: To 1000 ml double-distilled water (e.g. from a Millipore unit) in a baked glass flask add, under a fume hood, 1 ml DEPC (Sigma Chemicals, St Louis, MO). Stir for at least 30 min, until the oily substance is dispersed. Autoclave twice.
5× buffer DIG-I: To 121 g Tris–HCl and 87.6 g NaCl dissolved in 1260 ml double-distilled water add 740 ml 1 N HCl to pH 7.5. Autoclave. Dilute to 1× with distilled water. The 1× buffer does not need to be autoclaved.
1× TBS-Mg buffer pH 9–9.5: Bring 100 ml 1 M Tris–base to pH 9.5 with about 25 ml 1 N HCl, then add 20 ml 5 M NaCl to 950 ml double-distilled water. Add 50 ml 1 M $MgCl_2$. Mix and use.
MeHgOH solution: Mix in a fume hood 4.75 ml DEPC-treated water with 1 ml 10× borate buffer (final concentration 1×), 0.25 ml 1 M MeHgOH (Serva; final concentration 25 mM), 2 ml glycerol (final concentration 20%), 2 ml 1% bromophenol blue (final concentration 0.2%). Aliquot at 1 ml and store frozen. To use, mix 1 : 1 with the heat-denatured (10 min 65°C) RNA sample and load on an agarose/borate gel.
10× borate buffer: 30.09 g boric acid (0.5 M), 19.10 g sodium tetraborate × 10 H_2O (0.05 M), 14.2 g sodium sulfate (0.1 M) add 1000 ml double-distilled water. Add 1 ml DEPC, stir, autoclave.
Ethidium bromide stock: 10 mg/ml sterile water. Carcinogenic! Use 2 µl per 100 ml agarose solution.
20× SSC: 3 M NaCl, 0.3 M sodium citrate in double-distilled water. Treat with DEPC, autoclave twice.
EDTA: 0.5 M, dissolves when pH reaches 8.0. Treat with DEPC, autoclave twice. Make up small amount, aliquot and freeze.
LiCl: 4 M, treat with DEPC, autoclave twice. Make up small amount, aliquot and freeze.
Alkaline phosphatase substrates: NBT (nitroblue tetrazolium; Sigma) is 75 mg/ml in 70% (v/v) dimethylformamide. Store aliquots at –20°C. BCIP (X-phosphate, or 5-bromo-4-chloro-3-indolyl-phosphate toluidinium salt; Sigma) is 50 mg/ml in dimethylformamide. Store at –20°C. We usually have 2 ml of each in the freezer.
Other chromogenes for alkaline phosphatase: A red reaction product is yielded with Fast Violet B (final concentration 0.2 mg/ml) and naphthol-AS-E-phosphate (final concentration 0.1 mg/ml), both from Serva. The combination may be used for histological detection.

11.3 The Tissue

ISH on different kinds of "brain tissue" is presented in this chapter, namely rat brain, and organotypic cultures of rat visual cortex grown on coverslips with the roller-tube technique (Obst and Wahle, 1995). It also works in dissociated cultures of rat hypothalamic neurons (Wahle *et al.*, 1993).

11.3.1 Fixation by Perfusion or Immersion

Animal brain tissue is optimally fixed by vascular perfusion of a physiological salt solution followed by the fixative (Protocol 11.4; see also Ausubel *et al.*, 1987, chapter 14). The salt solution rinses out the blood so it does not clog blood vessels and allows the fixative to reach the capillaries.

The solutions are in glass bottles with an outlet placed about 1 m above the animal. The bottles are connected with tubing to a T-shaped adaptor. Stopcocks are placed in the tubing to start and stop the flow and to regulate the flow rate. The adaptor's third arm connects to the perfusion cannula with a piece of tubing that is smaller in diameter. Instead of a cannula, a piece of tubing (diameter approximately the size of the arterial stem) can be inserted into the heart. The cannula is filled to the tip with salt solution, but fixative should not be used in the first flush. Solutions are not autoclaved.

Fixatives: We use freshly made 4% paraformaldehyde (see Chapter 1, Protocol 1.2), which gives the best retention of cellular mRNA. Use ice-cold, if this is required for subsequent immunohistological procedures. Use additives, if required for detection of antigens, e.g. glutardialdehyde is required for detection of amino acid transmitters. The concentration should be kept as low as possible. The ISH procedure in sections tolerates 0.2% glutardialdehyde. Picric acid improves the fixation of very immature tissue; the ISH procedure tolerates 1% picric acid. Other fixatives have not been tried in our lab. It is not necessary to add DEPC. Never autoclave fixatives!

Tissue is dissected out of the skull, and is dropped into fixative. Gloves and clean equipment are used to handle the tissue. The block containing the structure of interest is dissected from the brain and is rinsed in fixative. With fine forceps the dura and pia mater are removed and the tissue block is then rinsed in several changes of fixative. The block may be stored for several hours at 4°C in fixative (4% paraformaldehyde) to postfix. Postfixation time depends on the quality of the perfusion. Some antigens require longer fixation, and this has to be determined empirically.

Protocol 11.4 Perfusion of Animals

1. Deeply anaesthetize the animal by intraperitoneal injection of pentobarbitone (60 mg/kg body weight is a lethal overdose).
2. Wait until reflexes and breathing have stopped (pinch the toes, or touch the cornea of the eye).
3. Open the thorax with large scissors (the lungs will collapse).
4. Carefully open the pericard (fine scissors), and steady the heart with a pair of forceps by gripping the right ventricle.
5. Poke a hole the size of the perfusion cannula into the left ventricle (use the sharp tip of the fine scissors to penetrate the muscle).
6. Take care to not disrupt the septum. Aim towards the left ventricle.
7. Carefully insert the cannula filled to the tip with 0.9% NaCl in 50 mM sodium phosphate buffer pH 7.5 or commercial Ringer's solution in distilled water.
8. Cut open the right atrium (avoid cutting other parts of the heart or arterial stem, and start the perfusion. Blood should start flowing out of the right atrium. There should be no leakage out of the left ventricle (backflow occurs when the cut is larger than the cannula) or out of the nostrils/snout (this happens when the septum is penetrated or perfusion pressure is too high; immediately reduce the flow rate).
9. Wait about half a minute (rat; wait longer if a bigger animal is perfused), until the solution flowing out of right atrium starts to become clear.
10. Close, with the stopcock, the flow of saline, and open the flow of fixative. The body will show a muscular tremor, the limbs will stretch, and the neck and limbs will stiffen when the fixative reaches the muscles. This indicates a good perfusion. Perfuse about 200–400 ml of fixative per rat.

Organotypic "roller-tube" cultures are prepared as described by Obst and Wahle (1995) (see also Chapter 8). To harvest cultures for ISH, medium is briefly rinsed off in sterile 100 mM sodium phosphate buffer pH 7.5, or with Hank's Balanced Salt Solution (Gibco). The culture is then fixed with cold 4% phosphate-buffered paraformaldehyde for 30–60 min on ice, and then rinsed with several changes of sterile phosphate buffer. The culture technique requires tissue embedding in a plasma–thrombin clot, which is carefully removed without destroying the tissue. Glutaldehyde is not used if whole-mount cultures are to be processed. Glutaldehyde-containing fixatives apparently cause the plasma–thrombin clot and/or the glia layer that surrounds the tissue to become impenetrable for probes.

Dissociated neuronal cultures prepared according to standard protocols are grown in 35-mm cell-culture dishes. Cultures are rinsed with sterile PBS (0.1 M sodium phosphate, 0.9% NaCl) and fixed with 4% buffered paraformaldehyde at 4°C for 30 min (see also Chapter 7).

11.3.2 Preparation of Brain Sections

After fixation, the tissue block has to be cryoprotected before freeze-cutting. For a rat brain 50 ml of cryoprotection solution are prepared in a sterile beaker or 50-ml Falcon tube. The tissue block is transferred into cryoprotection solution and left overnight at 4°C. Next morning the block will have sunk to the bottom of the beaker. This indicates a complete immersion with sucrose. The block can now be frozen.

The tissue block is frozen from base to top on dry ice in a small puddle of Tissue-Tek (OTC compound, Miles) or other type of cryo-mounting medium. A fresh bottle should be used and then reserved for freezing tissue for *in situ* hybridization.

While freezing, the block should be surrounded completely with the cryo-medium. Alternatively, the tissue can be frozen directly onto the chuck in a –40°C cryostat before being equilibrated to cutting temperature (–15°C to –20°C) for about 15 min (Chapter 2).

Meanwhile the equipment is set up to cut and collect the sections (see Chapter 2). A normal cryostat knife (C-knife) is used. This should be handled with gloves. Usually, knives are oiled after use to prevent corrosion. The knife is cleaned with ethanol, rinsed in distilled water and dried with a rinse in ethanol. When dry, it is inserted into the cryostat and allowed to cool. To transfer sections we use L-shaped glass hooks bent and polished over a Bunsen burner flame from Pasteur pipettes baked overnight at 180°C. The hooks are then wrapped in aluminium foil. Sterile baked glass Petri dishes are used to collect the cryostat sections. For a routine procedure the first dish can be filled with 2 × SSC to dissolve the cryo-medium and the second dish with a mixture of 2 × SSC and hybridization solution. This allows stepwise equilibration of sections to the hybridization solution. Sections are then transferred to the hybridization solution.

For the incubations we use 30-well glass plates (about 17 × 20 cm; the wells are 2.5 cm in diameter and 2 mm deep) covered with a glass lid. These are easily cleaned, wrapped in aluminium foil and baked at 180°C overnight, and can be reused many times. When large series of sections are to be processed we use cell-culture plates (e.g. 36-well plates) or pool several sections into small glass vials (2–3 ml in volume).

Sections are cut at a thickness of 20–60 μm at a low speed. Section thickness depends on the type of experiment and also on the quality of fixation. Well-fixed material, or older developmental stages, or adult animals, or material for double-labelling experiments are usually cut at 20–30 μm. Immature tissue (e.g. brain of newborn animals) is very fragile, and thicker sections survive the procedure much better. We have successfully hybridized sections as thick as 75 μm, and whole-mount cultures might be even thicker than that. Generally, the better the sections were fixed the better the results of the hybridization.

We have also tried successfully vibratome sections for semi-thin plastic embedding as described earlier (Wahle and Beckh, 1992). The equipment is cleaned with 1% SDS followed by several washes in sterile water. Ethanol is used to dry the equipment if necessary. The tissue is mounted on a sterile chuck with HistoAcryl or cyan-based glue. The sections are then cut under sterile phosphate buffer.

11.3.3 Tissue Culture Systems

Organotypic cultures are usually hybridized *in toto* (see Fig. 11.3D). Fixative is rinsed off in three changes (10 ml each) of phosphate buffer, then transferred to 2 × SSC. The coverslips are then removed from the culture tubes with blunt forceps. With a scalpel, the plasma clot is carefully scraped away as well as possible. The tissue is then taken from the coverslips with a scalpel or razor, and floated into 2 × SSC/prehybridization solution, and transferred with glass hooks to prehybridization solution in a glass well plate (see also Chapter 8).

Dissociated cultures are rinsed with 2 × SSC to remove the fixative, followed by a mixture of 2 × SSC/prehybridization mix, followed by 0.6 ml prehybridization mix per small dish. It is important to pipette slowly, however, as some cells will unavoidably be washed off. Cultures are prehybridized for 4 h at the desired temperature.

11.3.4 Storage of Tissue and Sections

Brain tissue blocks frozen and embedded as described above can be wrapped in aluminium foil and stored for months at –70°C. Before use, the block is removed from –70°C, the foil peeled away, and the tissue mounted onto a cryostat chuck and allowed to equilibrate to the cutting temperature in the cryostat (at least 30 min). A block under investigation can also be stored at –70°C and may be reused later. The block should be covered completely with cryo-medium; this is necessary to prevent drying of the tissue. It is then wrapped in foil and stored at –70°C.

Once cut, brain sections may be stored at –20°C in hybridization solution in round-bottomed 2-ml Eppendorf tubes (few small sections per tube) or in 15-ml polypropylene tubes (larger sections). For some experiments it is necessary to cut several series of alternating sections. One or two are hybridized immediately, the other series are well equilibrated to hybridization solution, transferred to appropriate tubes and stored at –20°C (the solution will not freeze). For hybridization, simply warm up a series of sections to room temperature, and transfer sections separately into well plates with fresh hybridization solution. In this way aliquots of sections can be maintained, either to be included in an experiment as positive controls, or to quickly test newly synthesized probes. Normal results have been obtained on material stored for up to four months so far. There is no loss of mRNA or antigenicity. A hybridization buffer contains more or less the same ingredients (just more expensive) as solutions used to infiltrate tissue for long-term storage at –20°C in a non-frozen state.

Organotypic cultures are also stored prehybridized in Eppendorf tubes at –20°C. This way cultures of different age *in vitro* or maintained under different culture conditions can be collected over a period of time to be hybridized concurrently later. This is more efficient and helps to reduce experimental variabiliy.

PROTOCOL 11.5 EQUIPMENT AND SOLUTIONS FOR HYBRIDIZATION

Equipment: Baked glass vials, glass well plates and Petri dishes, aluminium foil, cryostat, C-knife, 1% SDS, technical grade ethanol, Pasteur pipettes, Eppendorf pipettes, sterile tubes and tips, hybridization oven.
Reagents: Molecular biology (or highest purity) chemicals for the solutions.
Solutions:
Paraformaldehyde 4%: Dissolve 40 g in 300 ml distilled water brought to 70°C (chemical hood) with few drops of 1 N NaOH. The milky solution should clear immediately. Reduce heat. Buffer with sodium phosphate to pH 7.5 and 100 mM final concentration; it is best to add half a volume buffer from a 2× stock solution. Make up with distilled water to 1000 ml. Filter and cool to room temperature. Add additives.
Cryoprotection solution: 20–25% sucrose (RNase-free reagent) dissolved in DEPC-treated 100 mM sodium phosphate pH 7.5. Do not treat sucrose solution with DEPC and do not autoclave.
Pre-hybridization solution: 50% formamide (Fluka, or Baker, deionized), 250 µg/ml heat-denatured and sheared salmon sperm DNA, 100 µg/ml yeast tRNA, 0.05 M sodium phosphate buffer pH 6.5, 4 × SSC, 5% dextran sulfate, 0.02% bovine serum albumin, 0.02% polyvinylpyrrolidone, 0.02% ficoll 400. We usually make 500 ml or so, having the ingredients as stock solutions. Note that all powdery substances have to be dissolved before making the final mix. The mix is then divided into aliquots of 15 ml and stored at –20°C. To thaw, bring to 65°C and vortex to completely dissolve any whitish precipitate. The hybridization solution is the same mix plus the cRNA probe. When diluting the cRNA probe, vortex vigorously to assure complete mixing.

11.4 Troubleshooting Tissue Quality

A major reason for failure of the ISH procedure is bad tissue quality. There are several reasons why tissue may look bad. (1) The tissue is not adequately fixed. It needs some expertise to perfuse an animal in order to obtain an optimally fixed brain. Try to keep surgery time within a few minutes to get access to the cardiovascular system as fast as possible. It is best to start the flow of saline while the heart is still beating. Take care to remove air bubbles from the tubing as these may clog arteries. When dissecting the brain try not to squeeze the tissue. (2) When perfusion pressure is too high blood vessels may disrupt, and cell membranes too. The morphology is destroyed, and worse, cellular mRNA gets washed off or starts to degrade. Under the microscope, such tissue has a lot of holes, and the ISH reaction product – if present – is not confined to cell bodies. It may help to lower the fixative bottles, reduce the flow rate of the solutions or to try smaller perfusion cannulas. (3) Cryoprotection is not adequate. This results in freezing artefacts (shatter, cracks). If this occurs more time should be allowed to immerse the block in sucrose. It must be sunk to the bottom of the container (through at least 5 cm of sucrose cushion). The time will depend on the block size. (4) Freezing the tissue may cause artefacts. Excess sucrose should be blotted away and the block covered (on a piece of sterile aluminium foil) with Tissue-Tek. Then the block is mounted with its base in Tissue-Tek on a cryostat chuck and allow to freeze from bottom to top. When the block is covered completely with Tissue-Tek, it is important to avoid thawing/refreezing of the surface, which will give cracks in the sections. Air bubbles in the Tissue-Tek should be avoided. (5) Expertise in cryostat cutting is required (see Chapter 2). Artefacts are often breaks parallel to the knife, because tissue is too cold. Time should be allowed for equilibration to a higher temperature. We usually cut with a block temperature of –15°C to –20°C, and chamber temperature set to –30°C (cryostat 2800E, Leica, Hamburg, Germany). It is also important to cut at a low speed.

11.5 The *In Situ* Hybridization

11.5.1 Serial Dilutions and Application of Probe

After the sections have been cut and equilibrated to hybridization solution in the glass well plates, they are carefully flattened out to remove folds,

PROTOCOL 11.6 *IN SITU* HYBRIDIZATION

1. Thaw out the cRNA probes on ice or in hand.
2. Vortex probe and spin down.
3. Prepare desired dilutions in hybridization solution.
4. Vortex vigorously, incubate at 65°C for 5 min (optional), vortex again, and spin down.

The optimal dilution depends primarily of the titre of the probe. A transcription of 1 μg template DNA may yield about 10 μg riboprobe dissolved in 100 μl TER. When a stock concentration of 100 ng/μl is assumed, a dilution of 1 : 2000 means 5ng probe per 80 μl incubation. One transcription would then be sufficient for hybridization of 2000 tissue sections. If, on the Northern blot of the cRNA probe, a clear band was seen, start with a dilution of 1: 300 (Fig. 11.1, lanes 1, 2 and 4). If a strong band develops, start with 1: 500 (Fig. 11.1, lane 3). Very concentrated probes (Fig. 11.1, lane 5) can be diluted up to 1 : 2000. This probe yielded nice signals when diluted 1 : 5000. We have carried out a serial dilution (1:300–1:2000) with a riboprobe specific for neuropeptide YmRNA, expressed in a subset of interneurons, in whole-mount organotypic cultures of rat cerebral cortex. We found virtually no difference in specific labelling. The overall background, however, was becoming lower in sections incubated with increasing probe dilution. The optimal dilution thus has to be determined empirically. It also depends on probe size, how many cells express the target mRNA, how many copies are contained in each cell body, the quality of the tissue, and how stringent the hybridization and the washes can be performed. Too concentrated probes will yield a terrible background (they hybridize to every cell).

Under routine conditions, we then use the optimal dilution. When subsequent double-labelling experiments are to be carried out, however, we incubate alternate sections in three different dilutions (e.g. 1:500, 1:1000 and 1:1500). The lowest dilution is expected to reveal all mRNA-expressing neurons. However, the staining intensity may be too high in a number of neurons to perform an immunofluorescence, since too much accumulated reaction product may hinder antibody penetration or may quench the fluorescence. The colour development in the sections incubated with the higher probe dilution is carefully monitored. It may result in lower levels of reaction product (to develop in a given time), and could be stopped when a clear signal is obtained in the neuronal population of interest.

5. For hybridization, remove glass well plates from the oven, remove the lid.
6. With pipette tips remove the prehybridization solution, and apply the hybridization with diluted cRNA probes.
7. Take care to cover the sections completely from above and below. The sections must be freely floating in the solution.
8. Trapped air bubbles and wrinkles have to be carefully removed. Use yellow pipette tips for this purpose. Do not be afraid of manipulating sections this way. Well-fixed and well-cut sections are quite stable (try to disrupt one), but nevertheless you should avoid poking holes into the structures of

interest. When hybridization is performed in glass vials, pipette off the prehybridization solution and apply the probe. Mix probe solution and sections by swirling, check that sections are floating.

9. Cover plate with lid and place in 45–50°C oven. Seal vials with parafilm, then cover it with a piece of aluminium foil in case the parafilm cracks during hybridization.

---

wrinkles and trapped air bubbles. Care is taken to immerse sections completely during all steps of the procedure. Partial drying dramatically increases the background. The amount of hybridization solution necessary to cover a free-floating section depends on its size. For a coronal adult rat brain section we use about 80 µl. For a section through one hemisphere of an adult cat, 120–150 µl may be needed. The plate is covered with the lid and prehybridized for 3–4 h at 45–50°C in an oven (Protocol 11.6).

Hybridization times should be determined empirically. We usually prepare the sections on day 1, apply the probes in the evening, and hybridize for 12–16 h, that is basically overnight. Proceed with washing on day 2. Longer times may be necessary when low-level mRNAs are to be detected. A convenient timetable is to cut the sections on day 1, and store them overnight in prehybridization solution at –20°C, preferably already sorted into the well plates. Early morning on day 2 place the plates at 45–50°C for 4 h and by noon apply the probe. Proceed with washing around noon on day 3.

Sections or cultures stored in prehybridization solution can be started at any time. The tubes are warmed to room temperature and the desired number of sections removed with a sterile glass hook and spread into the glass well plates. The sections are prehybridized for another one or two hours. Then the probes applied.

11.5.2 Washing the Sections

After the hybridization, sterility is no longer necessary. From that point on we handle sections with histology equipment, that is artists' brushes, normal lab dishes simply cleaned in a dishwasher, and without gloves (Protocol 11.7).

11.5.3 Immunodetection of Hybrid Molecules

This procedure (Protocol 11.8) requires blocking, antibody incubation, washes and colour detection. Avoid drying of sections throughout the

PROTOCOL 11.7 WASHING SECTIONS

1. Prepare the washing solutions and equilibrate at hybridization temperature or slightly higher in a waterbath (up to 55°C; note that temperature has to be determined empirically).
2. Remove plates from the oven.
3. With a brush fish the sections out of the wells and place into insets/trays (see equipment) in 2 × SSC at room temperature.
4. Rinse briefly.
5. Transfer to pre-warmed 2 × SSC and wash for 15 min at 45–55°C.
6. Continue washing in pre-warmed 2 × SSC/50% formamide, 0.1 × SSC/50% formamide, and 0.1 × SSC; 15 min each wash. This means that sections have to remain for 15 min at the chosen temperature (controlled by digital thermometer)! Wash temperature and duration could be varied. Try higher temperatures (50–55°C) or prolonged washes background is too high even though probe is sufficiently diluted.
7. After the last wash, transfer to 1 × TBS or buffer DIG-I at room temperature and equilibrate in three changes for about 15 min.

procedure. The blocking reagent comes from Roche, but could be substituted with a 1–5% bovine serum albumin solution.

Note that the AP reaction products are formazan salts, which are not particularly stable in ethanol and xylene-based media. So do not try long-term storage of sections: the reaction product may degrade. As an alternative to xylene, water-soluble mounting media can be used for coverslipping immunofluorescent sections. However, the quality of the stain tends to decline. We therefore recommend analysis of the sections as soon as possible.

Results

Our published articles (Wahle *et al.,* 1994; Obst and Wahle, 1995; Gorba and Wahle, 1999) have extensively documented mRNA-expressing neurons at low power and higher magnification. Typical results are shown in Fig. 11.2. The mRNA-expressing neurons can be clearly identified at low power (Fig. 11.2A) and at higher magnification (Fig. 11.2B,C). Neurons of different size are labelled. Figure 11.2C shows developing rat cortex. Whereas some neurons are well differentiated and express glutamic acid decarboxylase (GAD) mRNA abundantly, others still display immature features and express less GAD mRNA.

PROTOCOL 11.8 DETECTION OF HYBRID MOLECULES

1. While sections are in the hot washes, thaw an aliquot of 5× blocking solution.
2. Dilute to 1× with DIG-I. Put at 65°C. Vortex to make sure it is completely dissolved. The solution remains turbid, but there should be no visible particles.
3. Carefully clean the glass well plate (dishwashing reagent, long rinse with distilled water, let dry).
4. With brushes, transfer sections from DIG-I into block solution in the glass well plate. Use about 150 µl per well.
5. Block for 1 h at room temperature.
6. Dilute antibody (alkaline phosphatase-labelled sheep-anti-DIG, $F(ab)_2$ fragments; Roche) 1 : 1500 or 1 : 2000 in DIG-I. Depending on antibody lot, other dilutions (1 : 5000) may be tried.
7. Pipette off block solution, rinse sections in the glass wells briefly in DIG-I.
8. Remove DIG-I and apply the diluted antibody to the sections.
9. Cover plate with the lid and incubate for 2 h at room temperature.
10. Transfer sections to DIG-I and rinse 3 × 15 min at room temperature.
11. Clean the glass well plate carefully (the substrate solution may precipitate if in contact with dirty glassware).
12. Equilibrate sections to pH 9.5 in two changes of TBS-Mg buffer pH 9.5 for 5 min.
13. In a 15-ml polypropylene tube (or clean glass vial) make up substrate solution (always fresh) in TBS-Mg buffer. The alkaline phosphatase substrates are nitroblue tetrazolium salt (NBT) and bromo-chloro-indolyl-phosphate (BCIP). Per 10 ml TBS-Mg buffer pH 9.5 use 45 µl NBT stock and 35 µl BCIP stock. Vortex to mix. The alkaline phosphatase can also be developed with substrates giving a red reaction product (Fast Violet B/Naphthol-AS-E-phosphate).
14. Transfer sections to the glass well plate and incubate in about 120 µl of substrate solution. Take care to completely immerse the sections. Incubate in the dark because the substrate solution is light sensitive. Sections will turn slightly blue/purple after some hours. The colour reaction should then be monitored once in a while by simply placing the glass plate on the stage of a microscope. With low magnification, specifically labelled cells (blue reaction product) should be visible. A good probe should give labelled cells within 4–5 h.
15. If it is late in the day, the sections could be stored in substrate solution overnight at 4°C. The reaction will proceed slower. Alternatively, the substrate solution can be diluted (try 1 : 1 dilution with TBS-Mg buffer). Check the sections next morning, and continue colour development at room temperature. Mount a section in Tris–Mg buffer, monitor when still wet with microscope; transfer back to substrate solution if staining is too weak.

16. Continue the colour reaction until sufficient reaction product has developed in the neurons of interest (check with microscope).
17. Stop the reaction by transferring the sections to Tris–HCl buffer pH 7.5. Lowering the pH and removal of Mg as co-factor inhibits the alkaline phosphatase.
18. Rinse 3 × 15 min in Tris–HCl buffer at room temperature.
19. Mount sections on gelatinized slides and air-dry.
20. Store dry in dark slide box.
21. Before analysis, clear by brief dips in ethanol and xylene, and coverslip with Eukitt or other mounting medium. Analyse immediately.

Entire sections can be printed onto paper directly with a photographic enlarger to chart mRNA-expressing neurons (this is the same technique as for printing Nissl stains described in Chapter 1, Fig. 1.19 and for printing Nissl stains of embryos, Chapter 3, Fig. 3.1B,C). Figure 11.3A shows hemispheres of rat cortex at two anterior–posterior levels. GAD mRNA-expressing neurons appear as white dots. In Fig. 11.3A, intensely labelled neurons are present in the septal and basal forebrain area, whereas the striatum expresses a lower intensity. In Fig. 11.3B, the reticular formation of the thalamus is intensely labelled, as is the amygdala complex and dorsal hypothalamus. The hippocampus contains many labelled cells intermingled in pyramidal and granule cell layers. Note that this section has handling artefacts. Hippocampus and striatum were disrupted from the corpus callosum. A closer view of striatal tissue is shown in Fig. 11.3C. Intensely and moderately β-preprotachykinin mRNA-expressing spiny stellate cells are shown.

The hypothalamus contains intensely GAD mRNA-expressing cells *in vivo* (Fig. 11.3B) and also *in vitro* (Fig. 11.3E; from Wahle *et al.*, 1993). Dissociated embryonic day-13 rat hypothalamus cells had been cultured for four weeks. GAD mRNA-expressing neurons (arrows) can be clearly delineated from unstained cells in this phase-contrast photomicrograph (unstained cells appear grey).

An organotypic culture of rat visual cortex, hybridized *in toto*, is shown in Fig. 11.3D. As *in vivo* (Fig. 11.3A,B), many GAD mRNA-expressing neurons can be identified, and different labelling intensities are present. GAD neurons occur in all layers, including layer I. Clusters of GAD neurons may appear in tissue areas not optimally flattened to a monolayer. ISH for NPY mRNA in these cultures results in a much smaller subset of labelled neurons (Fig. 11.4).

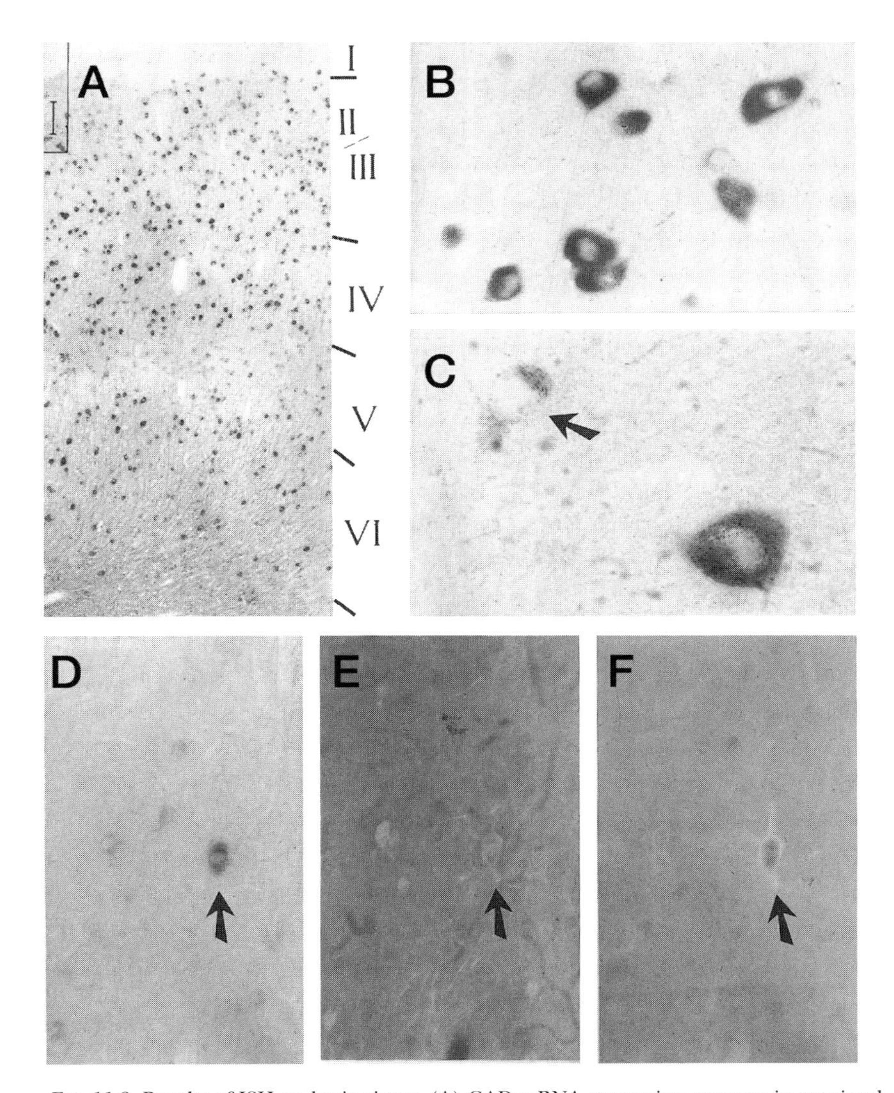

FIG. 11.2. Results of ISH on brain tissue. (A) GAD mRNA-expressing neurons in cat visual cortex occur in layers I–VI including the white matter below layer VI. (B) Intensely labelled cells at higher magnification. The cytoplasm contains the reaction product, the nucleus is free. (C) Besides a large intensely mRNA-expressing cell resides a small differentiating neuron which starts NPY mRNA expression in cat visual cortex. This finding corresponds well to immunohistochemical data (Wahle and Meyer, 1987). (D,E,F) Cat visual cortex supragranular layers. GAD mRNA-expressing neurons (D) co-express parvalbumin-immunoreactive material (E). The large soma (arrow), which intensely expresses GAD mRNA and also parvalbumin, is labelled by fluorescein isothiocyanate (FITC)-conjugated *Vicia villosa* lectin (F). These markers charactize the neurons of a large basket cell.

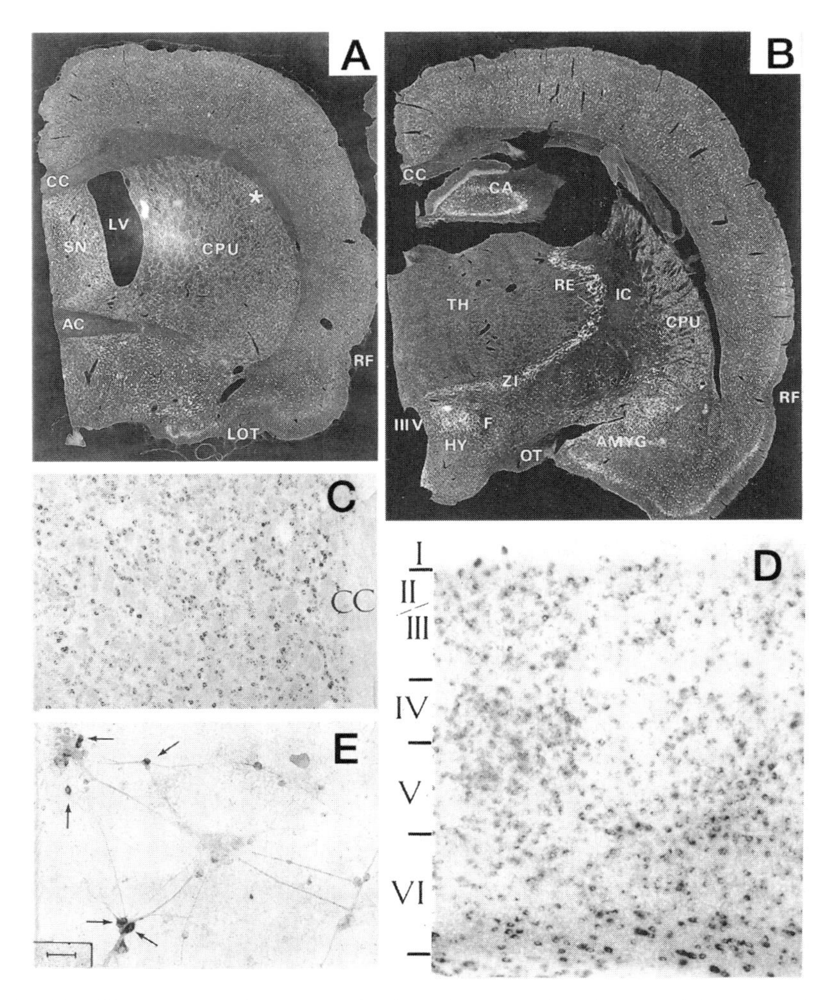

FIG. 11.3. ISH-stained sections printed with a photographic enlarger. (A) Frontal section through rat telencephalic centres. (B) Section through the diencephalon. The mRNA-expressing cells appear as white dots. Cell density and intensity of mRNA expression can be compared between the structures. (C) Higher magnification of rat striatum (corresponds to the area marked with an asterisk in (A)) in an alternating section revealing ß-preprotachykinin mRNA-expressing neurons. (D) GAD mRNA expression in neurons in organotypic cultures of rat visual cortex. Cells occur in all layers (indicated). (E) GAD mRNA expression in dissociated cultures (phase contrast). Arrows indicate GAD neurons. Other cell somata do not contain the dark-blue reaction product. AC, anterior commissure; AMYG, amygdala; CA, cornu ammonis of hippocampus; CC, corpus callosum; CPU, caudate putamen; IC, internal capsule; F, fornix; HY, hypothalamus; LOT, lateral olfactory tract; LV, lateral ventrical; OT, optic tract; RE, thalamic reticular formation; RF, rhinal fissure; SN, septal nuclei; TH, thalamus; ZI, zona incerta; III V, third ventricle.

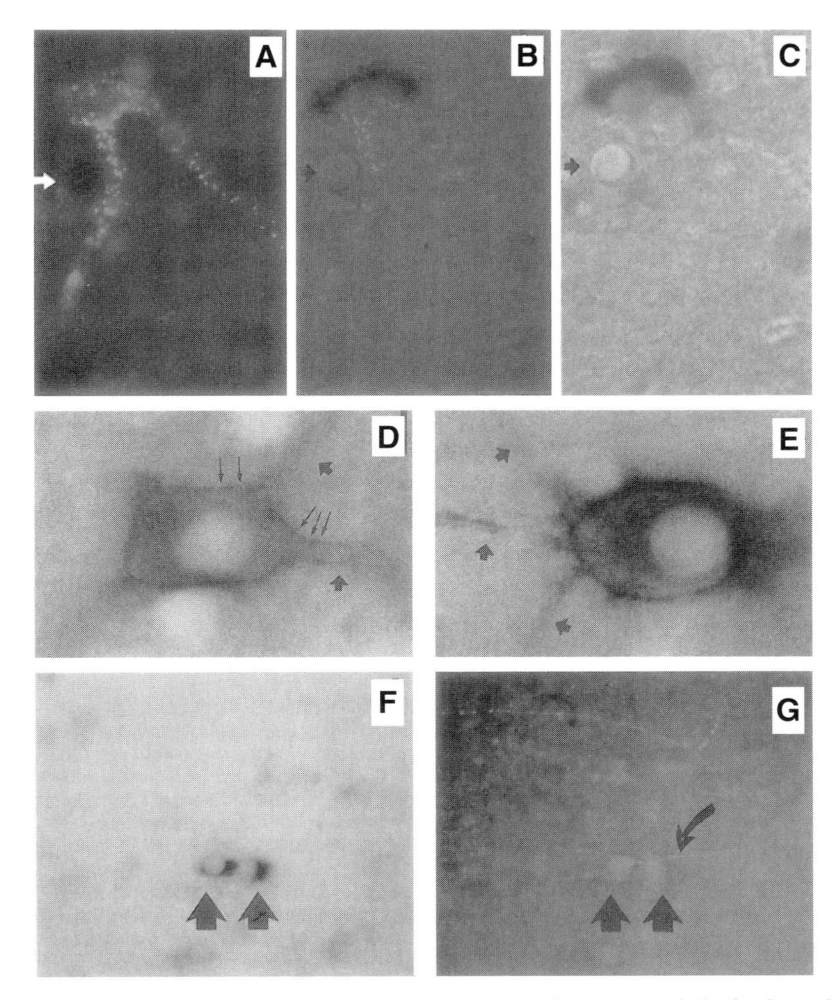

FIG. 11.4. Double-labelling experiments. (A) Retrogradely transported rhodamin-conjugated latex beads visualized with fluorescence have accumulated in the soma and dendrites of a neuron in the cat's nucleus of the optic tract. (B) GAD mRNA was visualized, and both products can be seen in a mixed exposure. (C) Bright-field exposure shows the AP reaction product in this neuron. (D) GAD mRNA, red reaction product, in a VVA lectin-labelled neuron. (E) Same as in D but blue reaction product. Neurons were from adult cat visual cortex. Fat arrows point to labelled processes; the fine arrows indicate the typical fenestrated VVA-labelling pattern. (F, G) Organotypic culture (explanted at postnatal day 2 and maintained for 2 weeks *in vitro*) of rat visual cortex displays NPY mRNA-expressing neurons (F, the bold arrows), which co-localize NPY-immunoreactive material, as revealed with FITC-labelled secondary antibody (G). The curved arrow points to the axon of one cell. Note the many immunoreactive beaded processes and puncta indicating the dense NPY innervation of the culture. This figure is reproduced in colour between pages 202 and 203.

11.5.4 Counterstaining the Sections

Since formazan reaction products are not stable in ethanol and xylene, the routine counterstaining, for instance with thionin (which stains the endoplasmatic reticulum), cannot be performed, because it requires prolonged exposure to differentiating ethanols. Also, the Nissl substance is where the AP reaction product has developed. Therefore, one would usually evaluate the ISH reaction first, and then counterstain selected sections, to obtain information about areal boundaries and cortical layers. Thionin will now stain preferentially the nuclei.

Another possibility is to counterstain with fluorescent dyes. The hybridized sections are immersed in DAPI (Sigma) 0.002% in phosphate buffer for several minutes followed by three washes in phosphate buffer. DAPI binds to the nuclear chromatin and gives a blue fluorescence under UV excitation. It allows one to distinguish neuronal from glial cells by nuclear morphology, and also gives a fair picture of cortical layers and CNS areas.

11.5.5 Troubleshooting the Staining Procedure

Besides probe and tissue quality we have had few problems obtaining good ISH signals. Optimal dilution is critical for a good signal-to-noise ratio, and has to be determined empirically. As Wisden and Morris point out in Chapter 1, most of the pretreatments are unnecessary for brain tissue. Only proteinase kinase (PK), by digesting proteins, enhances signal intensity due to exposure of target mRNA. All pretreatments, however, reduce tissue quality, and free-floating sections in particular tend to stick to brushes and equipment. Pretreatments may also destroy cell marker molecules or expose unwanted epitopes, which was our main reason for omitting these steps. RNases have so far not been a problem in ISH. After perfusion, the tissue is soaked with formaldehyde, which effectively denatures protein. The cryoprotection solution is made with DEPC-treated buffer and RNase-free sucrose, and sterile tools are used for handling. We use OTC compound (Tissue-Tek, Miles) for embedding. Equipment (cryostat, plexiglass blade, knife) is cleaned with SDS and ethanol, sterile tools are used for cutting, and some minutes after cutting the sections are in a formamide-containing solution, which again inhibits RNases. Post-hybridization washes and detection are carried out in non-sterile lab dishes. We have seen no necessity for RNase A post-treatment, since the reactions yielded a high specificity. The anti-DIG antibody is in a concentrated stable form; so far we have never had reason to believe that failure of an ISH has been due to failure of antibody/enzyme. The enzyme can be checked as follows: run the labelled control cRNA out on gel, and detect the band on the Northern blot. To check the enzyme, spot 0.5 μl antibody onto nitrocellulose and run an AP reaction. The blue spot indicates that the enzyme is working. It is important to ensure that the pH of the AP reaction is optimal, and that the substrates are mixed in the right concentrations.

RNase Contamination

Beginners are often afraid that poor signals are due to RNase contamination. However, endogeneous RNases are effectively killed by the fixation. Tongiorgi *et al.* (1998) recently found that only the pretreatment steps are sensitive to RNase contamination. The actual hybridization tolerates even exogenous RNases, provided the concentration remains below 100 ng/ml, which can hardly be achieved by accident.

In fact, together with Dr Gundela Meyer from University of La Laguna, I produced good hybridization results on immersion-fixed paraffin-embedded embryonic human cortex which was cut and slide-mounted under non-sterile conditions. The slides had been handled for weeks with bare hands. However, and for unknown reasons, in this case we found the minimalist hybridization buffer recommended by Wisden and Morris (Chapter 1, Protocol 1.8) much more effective.

Minimalist buffers or high-salt solutions are also found to be essential when DIG-labelled cRNA probes are used for Northern blot hybridization.

Sense Controls

Riboprobe transcriptions allow the transcription of the insert in the message sense orientation, yielding a cRNA identical to the cellular mRNA. In an ideal world these cRNA probes should not give any staining and thus may serve as "control". First of all, antisense and sense probes should be synthesized at equal concentrations (judge from the Northern blot). Then, serial dilutions should be run with both probes. With the probes used in our lab we find the following. Ideally, the sense probes do not yield any signals independent of dilution. Sometimes, however, the sense probes do yield signals, but these do not match the distribution of positively (antisense) labelled cells, or the intracellular staining pattern differs. We usually find that with increasing dilutions, the sense signals can be quickly diluted out, whereas the antisense signals gradually become fainter and at very high dilution detect only the strongest expressers, but still show the expected distribution of labelled cells. It is said by some that some sense sequences fail completely to yield negative staining. A reason may be that every sense probe is a sequence on its own and thus may find (partially) complementary mRNA sequences in the tissue. With regard to the high number of genes expressed in the brain this is not too surprising. If a sense control is required in such a case, it is best to subclone another fragment of the gene, either from coding or untranslated regions, and try again (see Chapter 1, Section 1.10 for more discussion on controls).

PROTOCOL 11.9 EQUIPMENT AND SOLUTIONS FOR WORKING AND DETECTION

Equipment: To rinse sections we use plastic staining trays with fitted Plexiglass plate insets which have regularly spaced holes (can be made by most machine shops). One side of the plate is covered with a nylon mesh glued to the Plexiglass. These plates are used for washing sections. Preferably, the matrix of holes in the Plexiglass plates is arranged such that it corresponds to the matrix of wells in the glass well plates of tissue-culture plates used for hybridization. Our glass well plates have 30 wells (6 by 5), the Plexiglass plates have 72 wells (6 by 12), so the sections of two glass well plates can be rinsed in one Plexiglass plate. Also needed are fine artists' brushes and gelatinized slides.
Reagents: As in Sections 11.2 and 11.3.
Solutions:
20× SSC (Section 11.3) is diluted to 2× and 0.1× with double-distilled water. Do not autoclave.
2× SSC/50% formamide: Mix 2 × SSC 1 : 1 with formamide (Fluka or Baker, straight from the bottle).
0.1× SSC/50% formamide: Mix 0.1 × SSC 1 : 1 with formamide.
10× Tris–HCl (0.5 M): 121 g Tris dissolved in 300 ml distilled water, brought to pH 7.5 with 740 ml 1 N HCl. Fill up to 2 l with distilled water.

11.6 Combining Techniques: Immunohistochemistry, Tract Tracing, Triple Labelling

11.6.1 Antibodies Compatible with the Protocol

Protocol 11.10 presents a short version of our routine immunohistochemistry protocol.

We have successfully tested a variety of antibodies (Gorba and Wahle, 1999) in combination with ISH: antibodies to the neuropeptides NPY, somatostatin, vasoactive intestinal polypeptide, structural proteins (microtubule-associated protein or neurofilaments), calcium-binding protein (parvalbumin, calbindin, calretinin) and others. Examples of double-labelled cells are shown in Fig. 11.4F,G. Here, NPY mRNA-expressing neurons in an organotypic culture of rat visual cortex co-express NPY-immunoreactive material. In Fig. 11.2D,E,F, triple labelling is shown. A large

PROTOCOL 11.10 IMMUNOHISTOCHEMISTRY

1. Stop the colour development in 1 × Tris–HCl buffer pH 7.5. Do not allow sections to dry; if possible, proceed directly with the immunohistochemistry.
2. Rinse 3 × 15 min in 1 × TBS at room temperature.
3. Block sections in 1–5% BSA in TBS for 30 min at room temperature. Alternatively, use normal animal serum for blocking (e.g. normal goat serum 3% in TBS). Make sure that further antibodies of the immune cascade do not cross-react with the blocking serum, or with the sheep-anti-DIG antibody.
4. Transfer sections to the primary antibody at working dilution (depends on antibody titre and antigen to be detected and has to be determined on control sections made from the same tissue block).
5. Incubate for 12–16 h or longer at 4°C (depends on the antigen, and has to determined on control sections).
6. Rinse sections 3 × 15 min in TBS.
7. Transfer to secondary antibody usually diluted according to the manufacturer's recommendation.
8. Incubate at room temperature for 2–4 h (or longer at 4°C).
9. Rinse sections 3 × 15 min in TBS.
10. If secondary antibody is conjugated to fluorochrome, mount sections and coverslip when still moist with appropriate mounting medium. Analyse immediately.
11. If an immunoperoxidase reaction is carried out, incubate in tertiary enzyme-tagged antibody complex or ABC reagent at working dilution for 2–4 h at room temperature.
12. Rinse sections 3 × 15 min in TBS.
13. Develop the HRP reaction product with diaminobenzidine (DAB) as chromogen. Make up 0.02–0.03% DAB in Tris–HCl pH 7.5. Dissolve completely. Note that benzidines are suspected carcinogens.
14. Pre-incubate sections in DAB solution for 5–10 min at room temperature (in the dark).
15. Add H_2O_2 as substrate for the peroxidase to a final concentration of 0.003%. Incubate for 5 min at room temperature. Sections turn light brown.
16. Stop the reaction by transferring sections to rinse buffer (Tris–HCl pH 7.5), rinse several times and mount on gelatinized slides. Air-dry sections. Coverslip with xylene-based or watery mounting medium and analyse.

basket neuron in the adult cat visual cortex was identified by expressing GAD mRNA (Fig. 11.2D), parvalbumin-immunoreactive material (Fig. 11.2E) and a lectin-binding site (Fig. 11.2F) on the cell's outer membrane (see Section 11.5.3).

So far, all antibodies that gave staining on control sections of a tissue block also gave staining of virtually equal intensity on sections processed for ISH. The initial perfusion fixation (Section 11.3.1) is the most important factor for antigen preservation in brain tissue. Most antigens currently under investigation are intracellular antigens which could be well fixed with a paraformaldehyde perfusion. We have not tested antibodies to fragile extracellular, membrane-bound epitopes (e.g. gangliosides). We do not know if such epitopes survive the ISH, but we know that lectin-binding sites, for instance, do.

Detection of antigens requiring a high percentage of glutaldehyde in the fixative (e.g. amino acid transmitters GABA or glutamate) is less compatible with the above ISH protocol, because the procedure tolerates only small amounts of glutaldehyde. A higher percentage of this fixative strongly cross-links the tissue such that probe penetration becomes less effective. Borohydrate or proteinase K pretreatment may be used to reduce the cross-linking bridges to overcome this effect, but the small amino acid transmitter molecules may be lost during these steps. Effects on mRNA retention have to be evaluated.

Secondary and tertiary antibodies depend on the type of labelling reaction being carried out. Usually fluorescence-tagged secondaries are diluted according to the manufacturer's recommendation. The optimal dilution (signal versus background) must be determined on the tissue of interest.

An indirect immunofluorescence is performed quickly, and can be analysed immediately. To increase signal intensity, we usually employ a two-step protocol with a biotinylated secondary followed by a fluorescent avidin conjugate (e.g. Texas Red-labelled avidin).

The ABC reaction is another procedure. The secondary is an affinity-purified antibody against the primary, and it is tagged with biotin. The ABC reagent is made from two components, avidin or streptavidin and biotin tagged with an enzyme, usually horseradish peroxidase (HRP). The two components are titred together such that the resulting complex has free binding sites for the biotin tagged to the secondary (just follow the manufacturer's recommendation).

The enzyme for the colorimetric detection is usually HRP, and the chromogen is DAB. HRP destroys peroxides such as H_2O_2 to H_2O and O_2. Emitted electrons cause the polymerization of the DAB to a brown product. The critical parameter is the concentration of the H_2O_2. It should be kept low (max. 0.005% final concentration); 150 μl per 200 ml DAB reaction volume should be added from a prediluted (1% in H_2O, always prepare fresh) stock slowly into the solution. Stir to mix. Too much H_2O_2 causes a high background. As with different substrates for the alkaline phosphatase, other chromogens in an HRP reaction give differently coloured reaction products. The DAB reaction product is brown and stable. Sections can be viewed with a light microscope, and structural details can be much more easily analysed than with fluorescence. If the reaction product is light brown, the darker blue of the AP reaction product is easily detectable. However, depending on the amount of antigen in the target cell, the somata may be dark brown. Then it becomes hard to detect a smaller amount of AP reaction product. Also, expensive colour pictures are needed for documentation, whereas fluorescence can be photographed on black and white film.

11.6.2 Histochemistry: Lectin Probes

Plant lectins are of interest for neurobiologists. Membrane proteins bind lectins via carbohydrate domains. The expression of membrane proteins and their post-translational modifications are often cell type-specific, and certain lectins stain exclusively distinct neuronal types, often with the selectivity of a monoclonal antibody. Lectins could therefore be used as selective markers for cell types (Protocols 11.11 and 11.12). Lectins are obtained conjugated to enzymes or fluorochromes. We have tried the lectin *Vicia villosa* agglutinin conjugated to HRP and to the fluorescent marker FITC in combination with ISH. VVA delineates specific neocortical interneuronal cell types, the large basket cells, neurogliform cells and chandelier cells. The lectin would thus allow the identification of these types in a population of GAD mRNA-expressing neurons.

Double-labelled neurons are shown in Fig. 11.4D,E. They can be identified by a brown reaction product distributed along the cell's outer membrane, often extending far into the dendritic compartment. In the VVA/FITC-labelled material, the cell membrane is fluorescent. In Fig. 11.2D–F triple labelling is shown. The GAD mRNA was detected in the cytoplasm of a large basket neuron, which was identified by the lectin-binding site (green FITC fluorescence on the membrane) and a calcium-binding protein, parvalbumin (detected in the cytoplasm and dendrites by Texas Red fluorescence). Besides information about a biochemical feature of the neurons (expression of a particular carbohydrate), information is also obtained on the geometry of the dendritic tree, which helps when comparing the mRNA-expressing neurons with cell types initially described by classical morphological techniques (Golgi impregnation). The presence of parvalbumin is related to a physiological feature: these neurons are fast-spiking cells and fire burst of action potentials (Naegele and Katz, 1990). We assume that other extracellular epitopes are equally stable throughout the ISH, and could be detected subsequently with lectin probes or with monoclonal antibodies.

AP can also be developed with substrates giving a red reaction product (Fast Violet B/Napthol-AS-E-phosphate, Section 11.2). This results in a red cytoplasmic stain and a brown membrane lining. Results are shown in Fig. 11.4D compared with the blue reaction product in Fig. 11.4E.

PROTOCOL 11.11 LECTIN HISTOCHEMISTRY

1. Stop the colour development of the ISH in Tris–HCl buffer pH 7.5.
2. Rinse sections 3 × 15 min in Tris–HCl.
3. Dilute VVA–lectin conjugated to HRP in Tris–HCl to 5 μg/ml and incubate the sections for 12 h at 4°C. Alternatively, use fluorescence-conjugated lectins.
4. Rinse sections in Tris–HCl.
5. Develop HRP activity with DAB and H_2O_2 (see above). Sections turn light brown.

PROTOCOL 11.12 EQUIPMENT AND SOLUTIONS FOR DAB STAINING

Equipment: As in previous sections.
Reagents: H_2O_2 (approx. 30% stock Perhydrol, Merck), DAB (Sigma).
Solutions:
DAB stock: In order to reduce exposure to powdered DAB, make up a 100× stock solution by dissolving 2 g DAB in 100 ml distilled water (fume hood, gloves). Store 1-ml aliquots at –20°C. It is stable for a year. Use one aliquot per 100 ml 1× Tris–HCl reaction volume. Lab dishes in contact with benzidines are immersed in hypochloride solution to bleach the DAB.

11.6.3 MULTI-LABELLING STRATEGIES

In a recent paper we have successfully performed a triple labelling by using a double *in situ* hybridization followed by an immunofluorescence (Fig. 11.5 – see Gorba and Wahle, 1999). The experiment was set up to determine whether parvalbumin-expressing basket and chandelier cells in the rat neocortex co-express the neurotrophin receptors trkC and trkB. This would tell us whether this neuron population could potentially respond to neurotrophin NT-3 or to the neurotrophins BDNF and NT-4/5, respectively.

One cRNA probe was synthesized with DIG-UTP and was against the neurotrophin receptor trkC mRNA. The other probe was made with biotin-UTP and was against the neurotrophin receptor trkB mRNA (for detailed background on biotinylation of nucleic acid to make probes, see Sambrook and

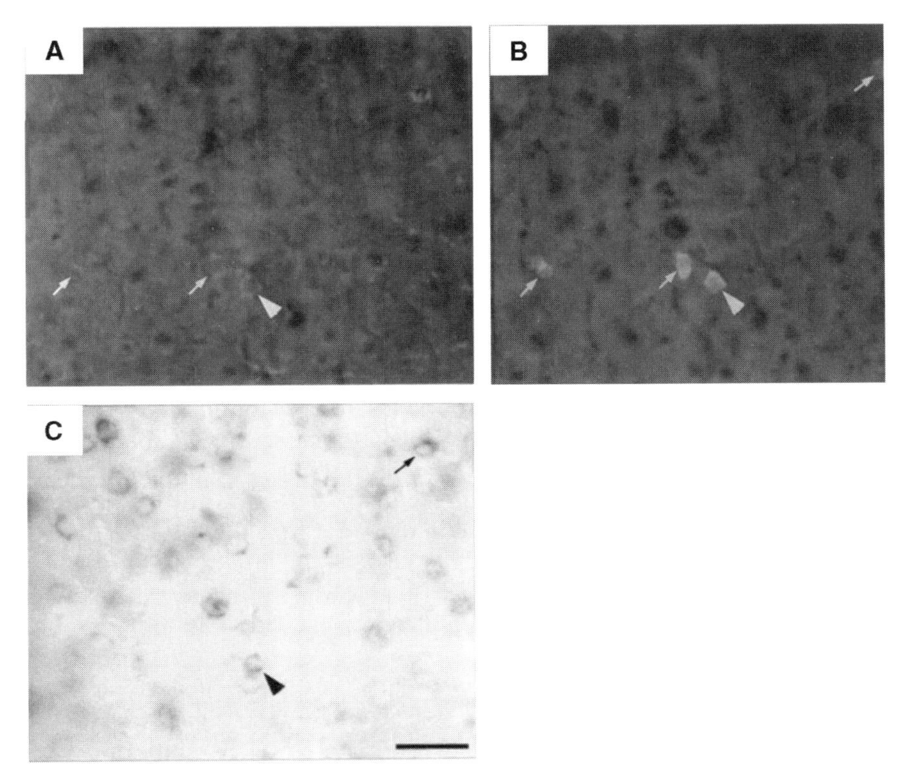

FIG. 11.5. Parvalbumin-immunoreactivity neurons expressing trkB or trkC mRNA are partially overlapping subsets. Triple labelling for trkB mRNA (A), parvalbumin immunofluorescence (B) and trkC mRNA (C) in visual cortex sections *in vivo*. Large arrowheads indicate a triple-labelled neuron, small arrows double-labelled neurons. Scale bar in C 50 mm for all photomicrographs. (Reproduced from Gorba and Wahle, 1999). This figure is reproduced in colour between pages 202 and 203.

Russell, 2001, section 11.116). (Biotin-UTP comes as 10 × RNA labelling mix from Boehringer; for making biotin-labelled cRNA probes, use Protocol 11.1, but substitute biotin-UTP for digoxigenin-UTP.) Both probes were checked on Northern blots, and used at about equal concentrations that were tested in separate *in situ* hybridizations for specific hybridization signals (the procedure for developing biotin probes is given at the end of this paragraph). The two cRNA probes were then mixed and hybridized together (at the same concentrations as used individually) to the sections of interest, followed by washes. After the blocking step, the AP-conjugated sheep-anti-DIG antibody (detection of the trkC mRNA) was applied together with a mouse monoclonal against biotin (detection of the trkB mRNA) and a rabbit antiserum against parvalbumin. (*Biotin detection*: probes tagged with biotin are detected with a cascade of mouse antibiotin (Dakopattas, Hamburg, Germany, dilution 1 : 500) for 12 h at 11°C, followed by biotinylated anti-mouse antibody (Dakopatts, dilution 1 : 300) for 3 h. The final reagent in this cascade, avidin–Texas Red (Serva, Heidelberg, Germany) 1 : 100 in TBS, is applied for 30–60 min after the development of the alkaline phosphatase reaction product. The OTC is rinsed, mounted and coverslipped with Immunofluore (ICN) (Gorba and Wahle, 1999).)

Now the trick: the biotinylated cRNA should be visualized as red fluorescence (Fig. 11.5A). The parvalbumin immunoreactivity should be visualized as green fluorescence (Fig. 11.5B). However, the pH of the AP reaction would destroy both fluorescences. We therefore reacted the secondary first: a biotin-conjugated goat-anti-mouse was used as secondary for the trkB detection cascade. Having these antibodies in place we developed the AP reaction product of the trkC cRNA to the desired intensity (Fig. 11.5C), and stopped the reaction. Now, as final reagents in the two other immune cascades we applied the Texas Red-conjugated avidin mixed with a FITC-conjugated goat-anti-rabbit.

The trkC mRNA-expressing neurons came with the blue AP reaction product, the trkB mRNA-expressing neurons came with a red fluorescence and the parvalbumin-immunoreactive neurons were in green. Quantifying the degree of overlap we found that about 80% of the parvalbumin neurons expressed trkB receptors and can thus be assumed to be dependent on the neurotrophins BDNF and NT-4/5. Only 12% of the parvalbumin cells expressed trkC mRNA, indicating that NT-3 is of less importance for the population of basket and chandelier neurons. Only a few parvalbumin neurons expressed both trkB and trkC mRNA and could potentially respond to both, NT-3 and BDNF or NT-4/5 (Gorba and Wahle, 1999).

11.6.4 Axonal Tract Tracing

The experiment described here is from the study by Wahle *et al.* (1994) (see also Schmidt *et al.*, 1998, 2001). We attempted to determine whether neurons in the nucleus of the optic tract (NOT) projecting to the lateral geniculate nucleus (LGN) express GAD mRNA. Previous attempts to immunohistochemically identify GAD or GABA had failed. For the surgery and injection protocols, the reader is referred to the original literature. We injected rhodamine-conjugated latex microspheres (beads; Luma Fluor New City, NY, USA) as a concentrated solution into the LGN. Three days were allowed for transport, during which the axon terminals of NOT neurons in the LGN took up the beads and transported them retrogradely towards the somata.

Perfusion and tissue processing followed Protocols 11.1–11.5. The midbrain containing the NOT was dissected and sectioned at 20-μm thickness from the posterior aspect of the LGN to mid-level superior colliculus. From each experiment about 180 cryostat sections were processed serially in well plates, hybridized to antisense GAD cRNA probe.

We identified two populations of GAD mRNA-expressing neurons. The small ones correspond well to interneurons of the NOT identified reproducibly with GABA immunohistochemistry. In addition, however, a population of large neurons was detected, expressing high amounts of GAD mRNA. These neurons had not been reliably observed with GABA immunohistochemistry. Quite a number of these neurons contained fluorescent beads (Fig. 11.4A–C). Thus, large pretecto-geniculate neurons express GAD mRNA and are thus GABAergic. The result unequivocally confirms their functional role as inhibitory neurons, which has been suggested on the basis of electrophysiology and morphology, but was under dispute, because attempts to localize the inhibitory transmittter GABA in the large neurons failed in most cases. ISH in this case has not simply revealed a neuronal mRNA; rather it has helped to define the role of a particular CNS cell type functionally modulating (via a disinhibition) the LGN relay neurons during saccadic eye movements.

Fluorescent beads are advantageous because they emit a strong fluorescence that is detected without any additional method (just the microscope). Nevertheless, we recommended monitoring the colour development of the ISH. If too much reaction product accumulates in the neurons, it may quench the fluorescence. Besides the beads, fluorogold has been used in combination with radioactive *in situ* hybridization (Burgunder and Young, 1988).

When enzyme tracers (e.g. HRP) are used, one must determine whether the activity survives the hybridization. The tracer detection should be carried out after the hybridization, because the other way round would mean having sections (with mRNA exposed) in a number of different (and hard to sterilize) substrate solutions and washes for quite some time. Lectin tracers could probably be used, since they can be detected with antibodies. Although we have not tried it, we assume they behave like other tissue antigens, which survive the hybridization.

11.7 Analysis and Presentation

The alkaline phosphatase reaction product developed with NBT and BCIP is amorphous and blue, or black in heavily mRNA-expressing cells.

The reaction product develops initially around the nucleus, where the concentration of cellular mRNA bound to ribosomes and endoplasmatic reticulum is highest. Later, more peripheral parts of the soma light up, and in several instances, e.g. with neuropeptide mRNAs, the staining extends for a considerable distance into the primary dendrites. If sections have to be analysed with high magnification, they have to be coverslipped with xylene-based media, and will soon be decayed. The water-based media allow observation up to 40× magnification, which is good for most purposes (e.g. analysis of double labelling). Under higher magnification the view is blurred.

Sections can be analysed with any normal light microsope. Nomarski optics (phase-contrast interference) could be applied. This visualizes, for instance, the nuclei of the labelled cells, and by nuclear morphology neurons can be distinguished from glia. Neuronal nuclei are larger, round or slightly ovally elongated, and have one round nucleolus. Glia nuclei are smaller, sometimes multiform and have one or more quite small nucleoli. When sections are counterstained with DNA-binding dyes (e.g. DAPI) the neuronal nuclei are faintly stained, and glial nuclei are more intensely stained due to a more condensed chromatin, and nucleoli appear as tiny blue dots under UV excitation. Since the nuclei are void of staining, no reaction product quenches the fluorescence.

Fluorescein, rhodamin and Texas Red – fluorochromes tagged to secondary antibodies or latex beads – are equally well detected in the somata and neurites (Fig. 11.4). In heavily labelled cells, the somatic immunofluorescence may be partly quenched by the reaction product, partly because we have frequently observed an inhomogeneous distribution of AP reaction product in the soma. Often it is more concentrated in one somatic pole. Therefore colour development must be stopped as soon as sufficient reaction product has developed in the soma. Then peripheral parts of the soma and the dendrites can usually be easily visualized by immunofluorescence, which in some cases is brilliant in the somatic pole opposite the mRNA-containing pole. This suggests that mRNA and peptide product are concentrated in different parts of a cell, a phenomenon explained by the known intracellular processing (translation, peptide trafficking and storage) pathways. Sometimes the nucleus appears labelled in immunohistochemically double-labelled cells because the fluorescence shines through the AP-negative nucleus from the cytoplasm behind. Double-labelled neurons may therefore display several patterns: fluorescence and mRNA either may co-extend in the same somatic parts, or may be accumulated in different somatic parts, or fluorescent neurites may arise from an mRNA-positive but fluorescence-negative soma. This accounts also for retrogradely labelled neurons. The beads and the mRNA may co-extend or may be localized in different compartments of the cell, e.g. beads in the dendrites, mRNA in the soma.

The same is true for the homogeneously brown DAB reaction product in immunoperoxidase reactions. Ideally, double-labelled neurons contain the blue-black amorphous AP reaction product around the nuclei and the brown DAB product in peripheral somatic parts or dendritic/axonal compartments of the neurons. Immunoperoxidase may be the method of choice when longer living preparations are preferred, or no fluorescence photomicroscope is available. It is also easier to reconstruct, for example, a peptide innervation pattern in the structure of interest, besides analysis of double-labelled cells. Furthermore, immunoperoxidase reaction products can be visualized at the electron-microscopic level. This allows the analysis of the synaptology of identified mRNA-expressing cells. Note, however, that it may be hard to distinguish dark-brown DAB from dark-blue AP reaction products.

The mRNA-expressing cells are photographed on colour or black and white film (e.g. Ilford FP4) or digitally and are presented as black and white prints on rather soft-grade paper for publication (see Chapter 1, Section 1.12). The distribution of mRNA-expressing cells can be charted with an X/Y plotter attached to the stage of the microscope or any computer-based reconstruction program. Alternatively, the sections (mounted on glass slides and coverslipped) may be placed into a photographic enlarger, and printed directly on black and white paper. This is a fast, convenient way to document mRNA-expressing neurons (shown in Fig. 11.3A,B). For double-labelled cells it is recommended to photograph any fluorescence first, followed by a mixed exposure (fluorescence excitation wavelength plus bright-field at very low intensity), then followed by the bright-field exposure.

Acknowledgements

Supported by DFG Neurovision.

References

Ausubel, F. M., Brent, R., Kingston, R. E., Moore, D. D., Seidman J. G., Smith, J. A., and Struhl, K. (Eds) (1987). "Current Protocols in Molecular Biology". Greene Publishing Associates and Wiley-Interscience, New York.

Burgunder, J.-M. and Young W. S., III (1988). *Mol. Brain Res.* **4,** 179–189.
Emson, P. C. (1993). *Trends Neurosci.* **16,** 9–16.
Gorba, T. and Wahle, P. (1999). *Eur. J. Neurosci.* **11,** 1179–1190.
Gorba, T., Klostermann, O., and Wahle, P. (1999). *Cerebral Cortex* **9,** 864–877.
Heumann, R., Goemans, C., Bartsch, D., Lingenhohl, K., Waldmeier, P. C., Hengerer, B., Allegrini, P. R., Schellander, K., Wagner, E. F., Arendt, T., Kamdem, R. H., Obst-Pernberg, K., Narz, F., Wahle, P., and Berns, H. (2000). *J. Cell Biol.* **151,** 1537–1548.
Naegele, J. R. and Katz, L. C. (1990). *J. Neurosci.* **10,** 540–557.
Obst, K. and Wahle, P. (1995). *Eur. J. Neurosci.* **7,** 2139–2158.
Obst, K. and Wahle, P. (1997). *Eur. J. Neurosci.* **9,** 2571—2580.
Obst, K., Bronzel, M., and Wahle, P. (1998). *Eur. J. Neurosci.* **10,** 1422–1428.
Sambrook, J. and Russell, D. W. (2001). "Molecular Cloning: A Laboratory Manual", third edition. Cold Spring Harbor Laboratory Press, New York.
Schmidt, M., van der Togt, C., Wahle, P., and Hoffmann, K.-P. (1998). *Eur. J. Neurosci.* **10,** 1533–1543
Schmidt, M., Sudkamp, S., and Wahle, P. (2001). *Exp. Brain Res.* **138,** 509–519.
Tongiorgi, E., Righi, M., and Cattaneo, A. (1998). *J. Neurosci. Methods* **85,** 129–139.
Wahle, P. and Beckh, S. (1992). *J. Neurosci. Methods* **41,** 153–166.
Wahle, P. and Meyer, G. (1987). *J. Comp. Neurol.* **261,** 165–193.
Wahle, P., Müller, T., and Swandulla, D. (1993). *Brain Res.* **661,** 37–45.
Wahle, P., Stuphorn, V., Schmidt, M., and Hoffmann, K.-P. (1994). *Eur. J. Neurosci.* **6,** 454–460.
Wahle, P., Gorba, T., Wirth, M. J., and Obst-Pernberg, K. (2000). *Development* **127,** 1943–1951.

CHAPTER 12

NON-RADIOACTIVE *IN SITU* HYBRIDIZATION: SIMPLIFIED PROCEDURES FOR USE IN WHOLE-MOUNTS OF MOUSE AND CHICK EMBRYOS

L. Ariza-McNaughton[1] and R. Krumlauf[2]

MRC National Institute for Medical Research, The Ridgeway, Mill Hill, London NW7 1AA, UK
[1]Present address: Vertebrate Development Laboratory, Cancer Research UK, PO Box 123, 44 Lincoln's Inn Fields, London WC2A 3PX, UK
[2]Present address: Stowers Institute for Medical Research, 1000 East 50th Street, Kansas City, Missouri 64110, USA

12.1 Introduction

The spatial distribution and timing of gene expression can be visualized using whole-mount *in situ* hybridization. This method was first used to detect gene expression in the *Drosophila* embryo (Tautz and Pfeifle, 1989). One of the main advantages is the ability to visualize the overall pattern of the spatial and temporal expression of a gene of interest in the whole embryo, a feature which makes whole-mount *in situ* hybridization an extremely powerful tool in developmental biology. In addition, the hapten-labelled probes used in this technique are safer and more stable than the probes used in radioactive *in situ* hybridization (see Chapter 11). In recent years the technique has been adapted to detect mRNA in many species (Wilkinson, 1995). We have used the technique extensively to determine the localization and patterns of expression of many genes expressed during development, but in particular for the expression of Hox genes in mouse, chick and zebrafish (Fig. 12.1). (Gavalas *et al.*, 2001; Manzanares *et al.*, 2001; Morrison *et al.*, 1997; Trainor *et al.*, 2002.)

INTERNATIONAL REVIEW OF NEUROBIOLOGY, VOL. 47
ISSN/02 $00.00

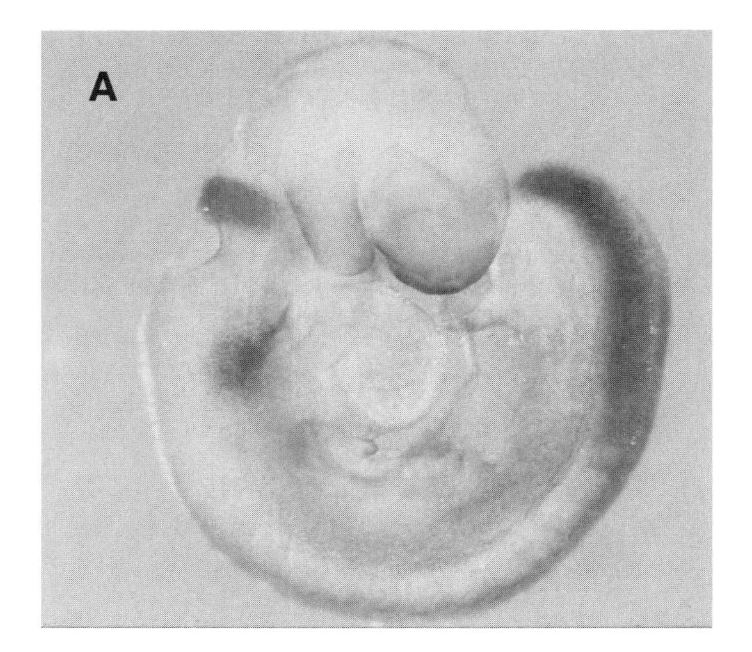

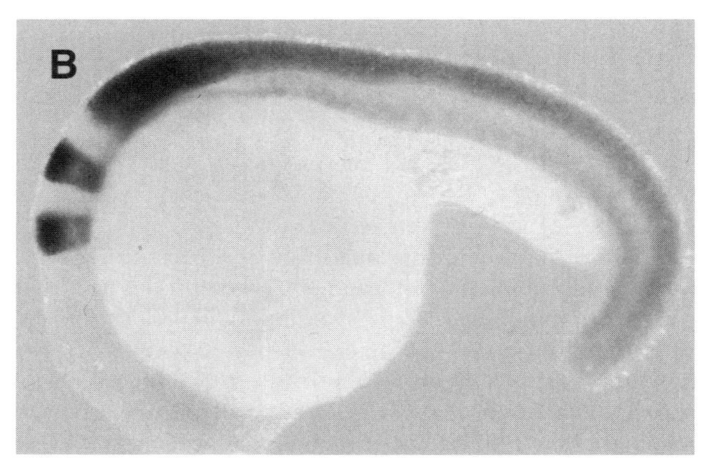

FIG 12.1. Patterns of expression of genes in two different species. (A) Whole-mount *in situ* hybridization of 9.5 dpc mouse embryo hybridized with a Hoxb1 RNA probe. There is a band of expression in the hindbrain corresponding to rhombomere 4. At this stage the expression in the neural tube has regressed posteriorly. (B) Zebrafish embryo (20 h), where two different probes have been used simultaneously. The two stripes correspond to Krox20, which is expressed in the hindbrain in rhombomeres 3 and 5. The most posterior staining in the neural tube corresponds to the pattern of expression of a presumptive Hoxa4 gene. This figure is reproduced in colour between pages 202 and 203.

12.2 Improved Whole-mount Methodology

Whole-mount *in situ* hybridization is a technique that it is not difficult as such, but it can be very frustrating at times, particularly when we first start learning. We have made a number of modifications to the basic method, with the aim of improving the reproducibility and reliability and of reducing the time to obtain results. The methods that we have found to be optimal are detailed below. Routine preparation methods for solutions are given in Section 12.8.

12.3 Probe Synthesis

A standard *in vitro* transcription reaction of the DNA template, incorporating digoxigenin-11-UTP, is carried out (see also Chapter 11, Section 11.2). The cDNA of interest is cloned into a plasmid vector, such as pBluescript (Stratagene) or pGEM (Promega), containing T7, T3 or SP6 RNA polymerase initiation sites. We have carried out hybridization using intact probes with a size range from 200 bp to 2 kb, with very satisfactory results. However, if probe penetration is a problem, the size can be reduced by partial alkaline hydrolysis to generate small probes of the order of 500 nucleotides (see Section 12.7).

12.4 Preparation of Tissue

One problem that we may encounter with whole-mount *in situ* hybridization is the trapping of reagents, for example in embryonic cavities, leading to problems of high background. The larger the embryo the more likely it is that there will be problems with high background and probe penetration. It is important to dissect and remove as much extra embryonic tissue as possible in order to increase access to target tissues (Morrison *et al.*, 1997). Structures such as the neural tube, heart, otic vesicles, etc. can trap reagents and therefore should be dissected open or punctured with a syringe needle to prevent non-specific trapping. You can do whole-mount *in situ* hybridizations on intact mouse embryos up until 10.5dpc; beyond this age, the method will only work with pieces of embryo. For chicks, the whole-mount

PROTOCOL 12.1 PROBE SYNTHESIS

To synthesize the antisense or sense probe, mix the following reagents at room temperature (detailed recipes for components are given in Section 12.8):

Sterile distilled water	10.0 μl
5× transcription buffer	4.0 μl
0.1 M DTT	2.0 μl
DIG 10× nucleotide mix	2.0 μl
Linearized plasmid (1μg/μl)	1.5 μl
RNasin ribonuclease inhibitor	0.5 μl
SP6, T7 or T3 RNA polymerase	1.5 μl

1. Incubate for 2 h or overnight at 37°C.
2. Remove 1μl aliquot and run on 1% agarose/TAE gel containing ethidium bromide.
3. Load the gel and run at 100 V for about 20 min. Estimate amount synthesized; an RNA band 10-fold or more intense than the plasmid should be seen.
4. Add 2 μl of DNase I (ribonuclease free) and incubate at 37°C for about 1 h to degrade plasmid DNA.
5. Add 50 μl distilled water, 25 μl 10 M ammonium acetate and 300 μl ethanol to precipitate the probe in order to remove non-incorporated nucleotides.
6. Mix and place in dry ice for 30 min or leave overnight at –20 or –70°C.
7. Spin in microfuge at 4°C, 30 min at 13 000 rpm.
8. Wash pellet with 70% ethanol. Remove and dry any residual ethanol in SpeedVac for approx. 3 min, or simply air-dry.
9. Redissolve in 50 μl DEPC-treated water. Can now be stored indefinitely at –20°C.
10. For hybridization use 2μl to each 1ml of hybridization mix.

method works up until stage 18; beyond stage 18, you can do whole-mounts on the body if you remove the leg and wingbuds. In our laboratory, we have used zebrafish embryos up to age 24 h post-fertilization (Fig. 12.1B).

Embryos can be processed either in small glass vials or in 1.5-ml Eppendorf tubes. Fixation in 4% paraformaldehyde (see details below) is followed by proteinase K treatment to increase probe penetration. The extent of this enzymatic treatment depends very much on the developmental stage of the embryo. It is important to optimize the timing as excessive digestion can cause tissue disintegration. On the other hand, poor signals are obtained if the embryo is not treated for long enough.

PROTOCOL 12.2 TISSUE PREPARATION

Unless otherwise stated all steps are carried out at room temperature.

1. Dissect out embryos in PBS, removing extra-embryonic membranes and any surrounding tissue that might interfere with the accessibility of the target tissues to the probe. Mouse embryo dissection is described in Chapter 3 (section 3.3).
2. Fix in 4% paraformaldehyde for 2 h at room temperature or overnight at 4°C. The time of fixation depends very much on the size of the embryo. This time works well in embryos of up to 9.5 dpc mouse or stage 16 chick embryo. Older embryos are fixed overnight.
3. Rinse embryos twice in PBT for 5 min.
4. Dehydrate embryos in a graded ethanol series diluted in PBT. We use 50% ethanol for 10 min initially, followed by two final rinses in 100% ethanol. Embryos can be stored in 100% ethanol at –20°C for several months. Embryos can be transported under 100% ethanol at room temperature; we have not experienced significant loss of signal.
5. Embryos are rehydrated by washing for 5 min each with 75% ethanol, 50% ethanol, 25% ethanol and finally in PBT.
6. Treat the embryos with proteinase K, 10 µg/µl, for an appropriate length of time (5–20 min, depending on the size of the embryo – a certain amount of trial and error is necessary here). The older the embryo the longer it is exposed to proteinase K.
7. Rinse in PBT, followed by fixation in 4% paraformaldehyde for 10 min.
8. Wash twice in PBT, and proceed straight to pre-hybridization.
9. Add the hybridization solution without the probe (see below), 0.5–1 ml depending on number of embryos, and incubate for between 2 h and overnight at 62°C. By doing a pre-hybridization we block non-specific binding sites.

We use a heater block, which holds 1.5- or 2-ml microcentrifuge tubes. The whole block is placed on its side on top of a shaking platform, with gentle rocking. There is on the market a Grant mini hybridization oven, which is more suitable and convenient than the heater block.

If wanted, embryos can be stored long term at –20°C in the hybridization solution. However, we have experienced some loss of signal after prolonged storage.

PROTOCOL 12.3 HYBRIDIZATION

1. Remove the pre-hybridization solution and replace it with pre-warmed hybridization mix containing the probe. Incubate at 62°C (overnight or over a weekend) with gentle rocking.
2. Wash with pre-warmed 2 × SSC + 0.1% Triton-X, twice for 30 min each at 62°C.
3. Wash with pre-warmed 0.2 × SSC + 0.1% Triton-X, twice for 30 min each at 62°C.
4. Rinse twice with KTBT for 5 min, at room temperature.
5. Block embryos by incubating with 25% lamb serum (heat treated) in KTBT for 2 h at room temperature. At the same time, in a different tube, the antibody should be incubated in the 25% lamb serum in KTBT for the same length of time as the embryos.
6. After blocking, incubate the embryos in 1 : 2000 dilution of anti-digoxigenin antibody in 25% lamb serum in KTBT. This incubation can be done at room temperature for a minimum of 4 h, or left either overnight or over a weekend at 4°C.
7. Wash embryos at least five times for 3 h each time, in KTBT at room temperature. If convenient, washes can also be done overnight at 4°C.
8. After the washes, incubate the embryos twice for 20 min in alkaline phosphatase buffer. Levamisol at a concentration of 1 mM, which inhibits most endogenous alkaline phosphatases, can be added to the buffer at this stage, but in our experience we have not found its use to be necessary.
9. The colour development is carried out in the dark. Replace the last wash with 10 ml of AP buffer containing 45 μl NBT (nitroblue tetrazolium, 75mg/ml in 70% dimethylformamide) and 35 μl BCIP (5-bromo-4-chloro-3-indolyl-phosphate, 50 mg/ml in 100% dimethylformamide).
10. The colour reaction should be monitored regularly. As soon as the desired signal is achieved the reaction should be stopped by washing several times in KTBT. Postfixation in 4% paraformaldehyde for 2 h at room temperature or overnight at 4°C stops the reaction and stabilizes the stain.
11. Store the embryos in KTBT at 4°C. For long-term storage it is better to keep the embryos in 4% paraformaldehyde at 4°C.

12.5 Hybridization

Many embryos from different stages can be processed simultaneously for each probe of interest. Hybridization of a sense strand probe should be included as a negative control for non-specific signal and overall level of background staining. This is particularly important if you are investigating the expression pattern of a novel transcript (see Chapter 1, Section 1.10, and Chapter 11, Section 11.5.5 for discussions on controls).

The embryos are incubated with the hybridization mix containing the DIG-labelled RNA probe, prepared as described above, and the sites to which the probe binds are then detected with anti-DIG antibody coupled to alkaline phosphatase. A coloured precipitate is subsequently produced at the site of gene expression when the embryo is incubated with the alkaline phosphatase substrates NBT and BCIP.

For hybridization the probe can be used in the range from 0.1 to 1 µg/µl. The length of time for the hybridization can be from overnight to over a weekend at 62°C. When high background is found to be a problem, the concentration of the probe should be decreased. High stringency washings can also achieve a reduction in background.

12.6 Processing for Histology

If more detailed analysis is needed the tissue or embryos can be processed for histological sectioning (Fig. 12.2). Paraffin sections are commonly used, but one of the disadvantages is the loss of signal due to the use of solvent involved in the procedure. The whole embryo can also be sectioned either using a cryostat (see Chapter 2 for cryostat sectioning of embryos) or vibrotome (Manzanares *et al.*, 2001). In our hands the use of vibratome has proved to be satisfactory and a good way to get a rapid answer.

In brief, the embryo or tissue is embedded in 20% gelatine dissolved in water and melted at 50°C. After orientation of the embryo, the gelatine is allowed to set at room temperature for about 20 min, and the whole block is then placed at 4°C for about 15–30 min to allow the gelatine to harden. Once the block is set it is kept in 4% paraformaldehyde at 4°C. Before sectioning, give a quick rinse in PBS and proceed to cut.

Another alternative to sectioning is simply to dissect out the tissue of interest and flat-mount it on a microscope slide (Fig. 12.3), ready to be photographed (Gavalas *et al.*, 2001).

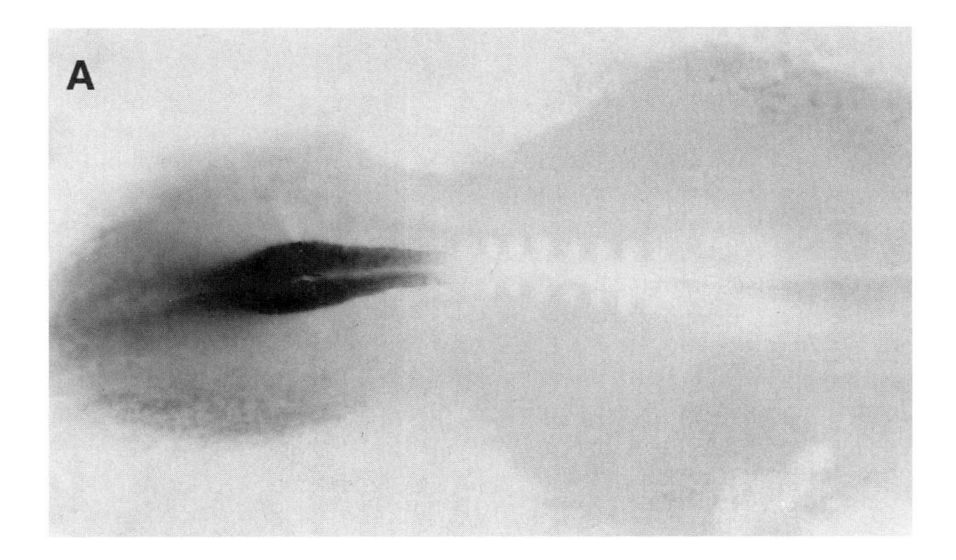

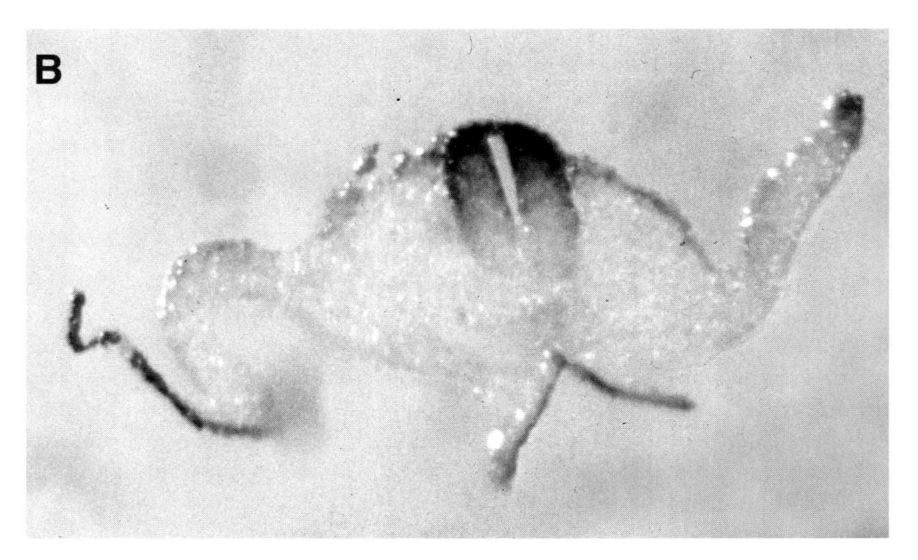

FIG. 12.2. Whole-mount *in situ* hybridization of a chick embryo. (A) Dorsal view of stage 8+ embryo, with chick Hoxb9 RNA probe. At this stage, there is a strong posterior expression in the presomitic mesoderm and neural tube. (B) Cryostat transverse section of similar embryo processed with the chick Hoxb9 RNA probe. Sectioning shows us that expression is stronger in the dorsal than in the ventral region. This figure is reproduced in colour between pages 202 and 203.

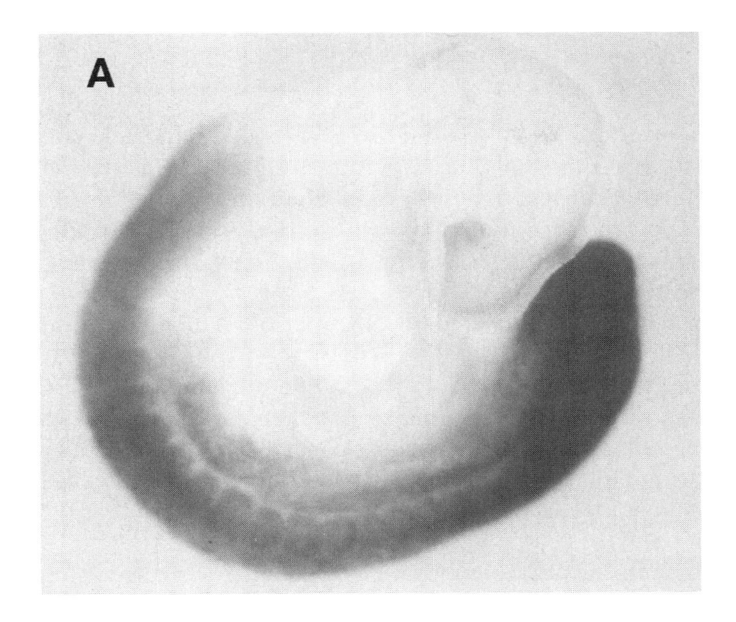

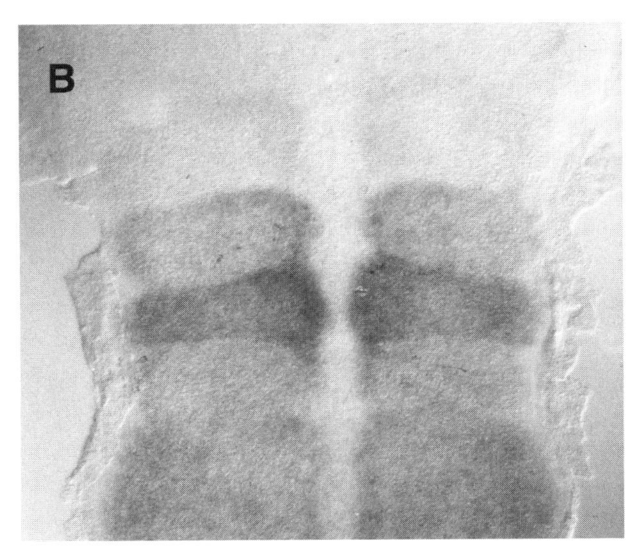

FIG. 12.3. Embryos hybridized with different hox genes. (A) 9.5 dpc mouse embryo hybridized with Hoxd4 gene, which is expressed along the neural tube and somites. Anterior expression is limited to rhombomere 6/7. (B) Flat-mounted hindbrain of a chick embryo stage 15, hybridized with the chick Hoxb2 probe. A clearer pattern of expression can be seen by dissecting out the tissue of interest and mounting it on a microscope slide. This figure is reproduced in colour between pages 202 and 203.

12.7 Notes

It is important to be patient and not to rush any step. Background is one of the main problems encountered when using these methods. To some extent reduction of background may also cause a reduction of signal, so there is a compromise between low background and signal/noise ratio.

To produce small-sized probes, partial alkaline hydrolysis can be carried out by adding equal volumes of 80 mM $NaHCO_3$ and 120 mM Na_2CO_3. These are mixed and heated at 60°C for a period of time, depending on the starting length of transcript. The size of the product is checked on an agarose gel, aiming for a size of around 500 nucleotides. Care should be taken as overdegraded probes can give low signals and high background.

Proteinase K treatment is important, particularly when dealing with large embryos where the accessibility of the probe can be a problem. Care is necessary, however, as overdigestion can result in low signal and damaged embryos.

Washing before colour developing is crucial. We usually wash for a minimum of 4 h, with several changes at room temperature or overnight at 4°C. After developing, the background can be reduced by washing with 0.5% Triton-X in KTBT.

12.8 Detailed Recipes

PBS

We use PBS (Dulbecco "A") tablets from Oxoid.

PBT

To 1 litre of PBS "A", add 1 ml of Triton X-100.

Hybridization mix (50% formamide/5 × SSC)

50 ml deionized formamide
25 ml of 20 × SSC
2 g Boehringer Blocking Powder
2 ml 10% Triton X-100
50 µg/ml Heparin
10 ml of 10 mg/ml tRNA
1 ml of 0.5 M EDTA
Make up to 100 ml with DEPC-treated water.

KTBT

50 mM Tris–HCl pH 7.5
150 mM NaCl
10 mM KCl
0.1% Triton X-100

Alkaline Phosphatase Buffer

10 ml Tris 1 M pH 9.5
5 ml $MgCl_2$ 1 M
2 ml NaCl 5 M
1 ml of 10% triton
20–25 mg Levamisole (optional)
Make up to 100 ml with distilled water.

4% Paraformaldehyde in PBS

Weigh 4 g of PFA in a fumehood. Add PBS "A" to make 100 ml. Place in a waterbath at 65–70°C. It will take 2–3 h to dissolve Cool on ice or overnight at 4°C (see also Chapter 1, Protocol 1.2, step 2).

References

Gavalas, A., Trainor, P., Ariza-McNaughton L., and Krumlauf, R. (2001). *Development* **128,** 3017–3027.

Manzanares, M., Bel-Vialar, S., Ariza-McNaughton, L., Ferretti, E., Marshall, H., Maconochie, M., Blasi, F., and Krumlauf, R. (2001). *Development,* **128,** 3595–3607.

Morrison, A., Ariza-McNaughton, L., Gould A., Featherstone, M., and Krumlauf, R. (1997). *Development* **124,** 3135–3146.

Tautz,D. and Pfeifle, C. (1989). *Chromosoma* **98,** 81–85.

Trainor, P., Ariza-McNaughton, L., and Krumlauf, R (2002). *Science* **295**, 1288–1291.

Wilkinson, D. G. (1995). *Curr. Opin. Biotechnol.* **6**(1), 20–23.

Appendix: Materials

	Material	Amount
1.	RNasin Ribonuclease Inhibitor Promega N2111	2 500 u
2.	T3 RNA Polymerase Promega P2083	1 000 u
3.	T7 RNA Polymerase Promega P2075	1 000 u
4.	SP6 RNA polymerase Promega P1085	1 000 u
5.	DIG, RNA labelling mix 10× concentrated Boehringer 1277073	40 μl
6.	Sheep serum Sigma S2263	
7.	Ribonucleic acid RNA, Torula Yeast Sigma R6625	100 g
8.	Heparin grade 1-A Sigma H-3393	100 000 u
9.	Anti-DIG-AP conjugate FAB fragment Boehringer 1093274	150 μl
10.	Triton X-100 Sigma T-8787	250 ml
11.	X phosphate solution BCIP Boehringer 1383221	3 ml
12.	(NBT) 4-Nitroblue tetrazolium chloride Boehringer 1383213	3 ml
13.	Formamide Boehringer 1814320	
14.	Blocking powder Boehringer 1096176	50 g
15.	Proteinase K (fungal) Boehringer 745723	100 mg
16.	Diethylpyrocarbonate (DEPC) Sigma D-5758	50 ml
17.	Levamisole Sigma L-9756	10 g

SUBJECT INDEX

Notes: Entries have been kept to a minimum under *In Situ Hybridization* and readers are advised to seek more specific topics.
Page numbers in *italics* denote figures/tables, page numbers in **bold** denote major entries.
Abbreviations used in subentries include:
AP: alkaline phosphatase
DIG: digoxigenin
ISH: *in situ* hybridization

Q

CONTENTS OF VOLUMES IN THIS SERIES

VOLUME 44

Volume 45

VOLUME 46

Neurosteroids and behaviour
S.R. Engel and K.A.Grant
Ethanol and Neurosteroid Interactions in Brain
A.L. Morrow, M.J. VanDoren, R. Fleming, and S. Penland
Preclinical Development of Neurosteroids as Neuroprotective Agents for the Treatment of Neurodegenerative Diseases
P.A.Lapchak and D.M.Araujo
Clinical Implications of Circulating Neurosteroids
A.R. Genazzani, P. Monteleone, M. Stomati, F. Bernardi, L. Cobellis, E. Casarosa, M. Luisi, and F. Petraglia
Neuroactive Steroids and CNS Disorders
M. Wang, T. Backstrom, I. Sundstrom, G. Wahlstrom, T. Olsson, D. Zhu, I-M. Johansson, I.Bjorn and M. Bixo
Neuroactive Steroids in Neuropsychopharmacology
R. Rupprecht and F. Holsboer
Neurosteroids and Human Disease
Lisa D. Griffin, Susan C. Conrad and Synthia H. Mellon